Multivariate Statistics - A practical
approach
B. Flury and H. Riedwyl

Practical Data Analysis for Designed
Experiments
B.S. Yandell

Practical Longitudinal Data Analysis
D.J. Hand and M. Crowder

Practical Statistics for Medical Research
D. G. Altman

Probability – Methods and measurement
A. O'Hagan

Problem Solving - A statistician's guide
Second edition
C. Chatfield

Randomization , Bootstrap and Monte
Carlo Methods in Biology
Second edition
B. F. J. Manly

Readings in Decision Analysis
S. French

Statistical Analysis of Reliability Data
M. J. Crowder, A. C. Kimber, T. J.
Sweeting and R. L. Smith

Statistical Methods for SPC and TQM
D. Bissell

Statistical Methods in Agriculture and
Experimental Biology
Second edition
R. Mead, R. N. Curnow and A. M. Hasted

Statistical Process control – Theory and
practice
Third edition
G. B. Wetherill and D. W. Brown

Statistical Theory
Fourth edition
B.W. Lindgren

Statistics for Accountants
S. Letchford

Statistics for Technology – A course in
applied statistics
Third edition
C. Chatfield

Statistics in Engineering – A practical
approach
AV. Metcalfe

Statistics in Research and Development
Second edition
R. Caulcutt

The Theory of Linear Models
B. Jorgensen

*Full information on the complete range of Chapman & Hall statistics books is available from the
publishers.*

Markov Chain Monte Carlo
Stochastic simulation for
Bayesian inference

Dani Gamerman

Professor of Statistics
Federal University of Rio de Janeiro
Brazil

CHAPMAN & HALL

London · Weinheim · New York · Tokyo · Melbourne · Madras

Published by Chapman & Hall, 2–6 Boundary Row, London SE1 8HN, UK

Chapman & Hall, 2–6 Boundary Row, London SE1 8HN, UK

Chapman & Hall GmbH, Pappelallee 3, 69469 Weinheim, Germany

Chapman & Hall USA, 115 Fifth Avenue, New York, NY 10003, USA

Chapman & Hall Japan, ITP-Japan, Kyowa Building, 3F, 2-2-1 Hirakawacho, Chiyoda-ku, Tokyo 102, Japan

Chapman & Hall Australia, 102 Dodds Street, South Melbourne, Victoria 3205, Australia

Chapman & Hall India, R. Seshadri, 32 Second Main Road, CIT East, Madras 600 035, India

First edition 1997

© 1997 Chapman & Hall

Printed in Great Britain by St Edmundsbury Press, Bury St Edmunds, Suffolk

ISBN 0 412 81820 5

A catalogue record for this book is available from the British Library

To André

Contents

Preface

The development of Statistics as an empirical science for data analysis and treatment has always been linked to the computational capabilities of the moment. It was not surprising that the widespread dissemination of faster computational equipment in the last decades enabled an unprecedented growth in the statistical treatment of complex models.

The area that benefitted most from this advance was applied Bayesian inference. Although the Bayesian approach has always had the support of many statistical users, its development has been hampered by the difficulty of its implementation in practical problems.

Recently, the story has definitely changed. The rediscovery of relatively simple but extremely powerful simulation techniques has enabled the application of the Bayesian paradigm to a variety of complex practical situations. And even better, these techniques are available for general use without the need for sophisticated pre-requisites of statistical theory for its understanding.

The aim of this book is to describe these techniques, collectively known as Markov chain Monte Carlo (MCMC, for short), to a wide-ranging public. The mathematical rigour used in the book is minimal and only requires the equivalent of a basic undergraduate course on Probability and Statistics. Anybody with that mathematical literacy can follow and understand all the material contained in this book. This includes not only statisticians but also mathematicians, operation researchers, engineers, economists and frequent users of statistics in general with some knowledge of the area.

The book grew out of lecture notes in Portuguese prepared for a short course on the topic taught at the XII Meeting of Brazilian Statisticians and Probabilists, held in Caxambu (MG) in August 1996. I tried to preserve the didactic character of the text in the expansion/translation. The book can be used as an advanced undergraduate course or graduate course on MCMC after at least a Probability course. With that aim, the book includes a set of exercises at the end of each chapter. Some of them have a more mathematical flavour and are directed to graduate students.

Previous notions of simulation, Bayesian inference and Markov chain are useful but are not needed. The first chapters of this book present a basic introduction to these topics. The presentation is geared towards applying them in the understanding of general MCMC methods. Other approximating techniques used in Bayesian inference are also described. They provide a useful benchmark for comparison with and the introduction of MCMC. The material covered in these chapters prepares the reader for later chapters and therefore many relevant points are made throughout these chapters. The variety of topics required for full appreciation of MCMC techniques has led to the inclusion of an introductory chapter at the beginning of the book. This chapter describes without technical details the route I chose to present the subject and serves as a guide to the book for the less mathematically oriented reader. It also describes the notation used in the book.

The core of the book consists of chapters 5 and 6 describing Gibbs sampling and Metropolis-Hastings algorithms respectively. These are the main MCMC methods used nowadays. Examples of inference using these methods are provided with emphasis on hierarchical models and dynamic models, previously introduced in chapter 2. Discussions over implementation, limitations and convergence issues are also included in these chapters. Finally, chapter 7 discusses further topics related to MCMC and more advanced material.

The book is an attempt to provide a thorough and concise presentation of the subject starting virtually from scratch. The conciseness precludes a very detailed coverage of all topics. Important theoretical and methodological points could not be skipped and practical data analysis was restricted to some of my own limited experience in the area. I concede that more could have been done but the extension would have required a substantial enlargement (and delay) of the book thus removing the slim (and updated) nature with which it was conceived. Chapman & Hall has recently released three excellent books (Carlin and Louis, 1996; Gelman, Carlin, Stern and Rubin, 1995; Gilks, Richardson and Spiegelhalter, 1996) dealing with MCMC theory, methodology and applications at a variety of levels that can be used to supplement the study of the subject.

The possibilities of expansion of MCMC methods are inumerous. Most of the relevant literature was published in the 90s and is providing a fascinating and profitable interaction between statisticians, probabilists, mathematicians and physicists. The area should still experience a considerable growth in the future both in theory and in implementation and applications level. This text provides an account of the current situation in simple terms so that its readers will also be able to take part in this process.

It is fair to say that MCMC has recently developed mostly with applications to Bayesian inference as inspiration, but nothing prevents their use on problems in frequentist inference. In essence, MCMC deals with drawing random values from a given distribution. The context that gave rise to the distribution is not mathematically relevant in general. The presentation however remains faithful to the main motivation that fuelled the use of MCMC techniques and will retain a strong Bayesian flavour.

My introduction to the area was eased by constant interactions with

Adrian Smith and Jon Wakefield during my visit to their department at Imperial College in 1994. My view of the subject and the preparation of this text bear their influence but I assume full responsibility for all views expressed here. Also, most of chapter 4 in the Portuguese version of this book was written by Tereza Benezath, a colleague at my department. The data analysis illustrating convergence diagnostics was performed by Aparecida Souza. The original idea of preparation of notes on the subject was given by Helio S. Migon, also from my department and coauthor of many joint research papers. Bent Jørgensen, Pablo Ferrari, Wally Gilks, Peter Green and Maria Eulalia Vares were also helpful with suggestions and encouragment. Mark Pollard, Stephanie Harding, and later Richard Whitby, David Hemsley and James Rabson provided good editorial support from Chapman & Hall's office. I find it important also to acknowledge the support provided by my University and Department through the use of their facilities and by the research supporting agencies CAPES, CNPq and FAPERJ through grants that enabled me to continue my research projects. Finally, I would like to thank my family, relatives, work colleagues and friends in general for the support during the difficult period of preparation of this book.

Dani Gamerman

Rio de Janeiro, July 1997

Introduction

Overview of Bayesian Inference

This is a book about Statistical Inference, the area of Science devoted to drawing conclusions or inference about data through quantitative measurements. There is often uncertainty associated with measurements, either because they are made with imprecise devices or because the process under which these quantifications become available is not entirely controlled or understood. The tool used to quantify uncertainties is probability theory and probability distributions are associated with uncertain measurements. The specification of probability distributions to the uncertain measurements or random variables in a given problem along with possible deterministic relations between some of them defines a statistical model.

An example considered later in the book (see Example 6.4) is the study of the impact advertising expenditure may have on sales of a product or some surrogate measurement such as advertising awareness. In this case, the uncertain measurements are the results from weekly opinion polls carried out in a given population of interest. The result of an opinion poll is given by the percentage y of people who remembered having watched the advertisement on TV. It is expected that advertising expenditure x might have an effect on awareness. Therefore a deterministic relation is established to link its effect on the awareness probability π. The simplest link is given by the linear relation $\pi = \alpha + \beta x^*$. Since $\pi \in [0, 1]$, it is usual in such cases to transform it to the real line before equating it to the linear form. A very common transformation is

$$\text{logit}(\pi) = \log\left(\frac{\pi}{1-\pi}\right) = \alpha + \beta x$$

In either case, the larger the value of β, the more effective the advertisement campaign is in boosting awareness. There are, of course, many possible relations that can be entertained. The collection of the possible relations along with probability specifications for the percentages from the polls defines a statistical model.

Once a model is built, there are many ways to proceed with inference. The Bayesian approach considers uncertainties associated with all unknown

---

* The value of the expenditure should take into account the instantaneous expenses but also the downweighted values of the expenses over previous weeks.

quantities whether they are observed or unobserved. Inference is drawn by constructing the joint probability distribution of all unobserved quantities based on *all* that is known about them. This knowledge incorporates previous information about the phenomena under study and is also based on values of observed quantities, when they are available. This book assumes the general case where both pieces of information are available.

In this case, the distribution of *unknowns* given the *knowns* is called the posterior distribution because it is obtained *after* the data is observed. The unknown quantities may include future observations (that are currently unknown). Inference about them is referred to as prediction and their marginal distribution is referred to as the predictive distribution. The operations required to obtain these distributions are derived in Chapter 2 and exemplified in a number of typical situations that occur in practice.

For the example, the quantities of interest are α and β, used to define a link between expenditure and awareness. They are unknown, otherwise there would be little point in performing the polls. Marketing experience may provide some background information on them. Another source of information is provided by the result of the polls once they are carried out and the percentages become known. Bayesian inference provides the tools to combine these pieces of information to obtain the posterior distribution of α and β based on previous, background knowledge and the observed information from the polls. There may also be interest in predicting future results from the polls for an anticipated advertisement expenditure in future weeks. In this case, the future results are added to the set of unknown quantities of interest.

Obtaining the posterior distribution is an important step but not the final one. One must be able to extract meaningful information from this distribution and translate it in terms of its impact on the study. This is mainly concerned with evaluation of point summaries such as mean, median or mode, or interval summaries given by probability intervals. In a few examples, this extraction or summarization exercise can be performed analytically, which means that an exact appraisal of the situation can be made. These cases are also illustrated in Chapter 2 and for them, the inferential task is completed.

In the advertisement study, the main interest is the evaluation of the value of β. If its distribution is concentrated with large probability around positive values then the study confirms that advertisement boosts awareness. Quantification is also important: the larger the values of β, the better the advertisement campaign is in raising awareness about the product.

In most cases, however, the complexity of the model prevents this simple operation from taking place. The complexity is sometimes caused by the combination of the sources of information available for a given quantity. In other cases, it is caused by the sheer amount of quantities required for an adequate description of the phenomena studied. In some cases, it may even be caused by a combination of many quantities with many sources of information for some of them.

In the example, a more adequate description of the process is provided by a model that allows the links between expenditure and awareness to change with time. Different habits, changing environment, other rival advertisement campaigns and change in advertisement campaign are all reasons for a dynamic modelling. One possible representation is to allow the quantities α and β to vary as time passes. The relation between the awareness probability π_t and the expenditure x_t at week t becomes

$$\text{logit}(\pi_t) = \alpha_t + \beta_t x_t$$

where the unknown quantities α_t and β_t are now allowed to change with the week. This automatically leads to a substantial increase in the number of the unknown quantities. Also, one must expect some degree of similarity between links in adjacent weeks. One convenient form to specify similarities is

$$\begin{aligned} \alpha_t &= \alpha_{t-1} + w_{1t} \\ \beta_t &= \beta_{t-1} + w_{2t} \end{aligned}$$

Note that the number of unknown quantities has risen dramatically from 2 to $2n$ where n is the number of weeks considered in the study. The incorporation of these similarities in the model also means that the structure of the model has increased in complexity. The distribution of the unknown quantities has consequently become more complex to handle.

One is inevitably led to seek approximations that can provide at least a rough guide to the exact but unobtainable answer. There are many ways to tackle this problem and a variety of suggestions have been proposed in the literature, with more emphasis on this aspect from the 80s. The timing is related to increased computing power enabling more sophisticated and computationally-based solutions. These solutions can be broadly divided into two groups: deterministic and stochastic approximations.

Some of the deterministic methods are based on analytical approximations whereas others are based on numerical approximations. They have received a great deal of attention in the literature and have been apllied with success in problems where the number of unknown quantities is small. Chapter 3 reviews the main approximating techniques, pointing at their strengths and weaknesses.

An entirely different perspective to extracting relevant information contained in a given distribution is provided by stochastic simulation. The approach here is to use values simulated from the distribution of interest. A collection of these values forms a sample and defines a discrete distribution concentrated on the sample values. The distribution of these values is an approximation to the parent distribution used for the simulation. Then, all relevant calculations with the parent distribution can be approximately made with the sample distribution. In particular, the sample can be grouped into intervals and the histogram of relative frequencies plotted. If a large number of these values is simulated then the resulting histogram will be a very close approximation to the density of the distribution of interest. Chapter 3 also describes approximating techniques based on stochastic simulation.

Stochastic simulation, or Monte Carlo, techniques have a few attractive features that may explain their recent success in Statistical Inference. First, they have strong support in probability results such as the law of large numbers (equation (3.8)). It ensures that the approximation becomes increasingly better as the number of simulated values increases. This number is controlled by the researcher and only time and cost considerations may prevent a virtually error-free approximation. Also, at any stage of the simulation process, the approximation error may be probabilistically measured using the central limit theorem (equation (3.7)).

The main thrust of the book is the description of techniques devoted to perform Bayesian inference based on stochastic simulation, hence its subtitle *Stochastic simulation for Bayesian inference*. Before applying simulation, it is important to present basic, direct simulation operations to those not familiar with them. This is the purpose of Chapter 1. Many of the results presented there will be returned to in a more elaborate setting in later chapters.

Using these techniques, it is possible to devise simulation schemes to draw values from the distribution of α and β (in the static model setting) but they do not provide adequate solutions to the more elaborate case of time-varying α_t and β_t. These techniques will tend to be very inefficient as the dimension of unknown quantities increase, and more sophisticated simulation techniques will have to be used.

Overview of MCMC

Nowadays, there are many problems of interest that fall into the category of large dimension models. Dynamic settings are just an example. Other examples also arise in the context of hierarchical or random effects models and models for spatial data. The first group roughly deals with unstructured additional variation whereas the second group deals with variations due to a neighbouring structure. They will also be considered in later chapters. Models with measurement errors and a mixture or combination of models are also settings for large dimension models.

The title of the book, *Markov Chain Monte Carlo*, refers to an area of Statistics, usually refered to as MCMC by taking the first letters of each word. MCMC will be described in detail in this book. It provides an answer to the difficult problem of simulation from the highly dimensional distribution of the unknown quantities that appear in complex models.

In very broad terms, Markov chains are processes describing trajectories where successive quantities are described probabilistically according to the value of their immediate predecessors. In many cases, these processes tend to an equilibrium and the limiting quantities follow an invariant distribution. MCMC techniques enable simulation from a distribution by embedding it as a limiting distribution of a Markov chain and simulating from the chain until it approaches equilibrium. Before understanding simulation through Markov chains, or MCMC in short, it is important that properties of Markov chains are well understood. For the sake of those not familiar with them, Chapter 4 reviews the most relevant results.

The introduction of Markov chains in the simulation schemes is vital. It allows handling of complicated distributions such as those arising in the large dimension models mentioned above. It is interesting that introduction of an additional structure, the Markov chain, into an already complex problem ends up solving it! There is also the matter of the extra work involved in simulation of a single value by MCMC: a complete sequence of values of a chain until it reaches equilibrium is required and only the equilibrium value can be taken as a simulated value from the limiting distribution. Fortunately, there are also analogues of the law of large numbers and central limit theorems (equations (4.6) and (4.9), respectively) for Markov chains. They ensure that most simulated values from a chain can be used to provide information about the distribution of interest.

There is still the question of how to build a Markov chain whose limiting distribution is exactly the distribution of interest, namely the distribution of all the unknown quantities of the model. It is amazing that not only is this possible but that there are large classes of schemes that provide these answers. One such scheme is Gibbs sampling. It is based on a Markov chain whose dependence on the predecessor is governed by the conditional distributions that arise from the model. It so happens that many models have a complex joint distribution but by construction (some of) their conditional distributions are relatively simple. Gibbs sampling explores this point and is able to provide simple solutions to many problems. Gibbs sampling is presented in Chapter 5 and exemplified in a number of situations including models with hierarchical structure and models with a dynamic setting.

There are many ways that MCMC can be used in any given situation. The main concern is efficent computation. Efficiency can be measured by the ease with which a simulated sample is obtained. It takes many aspects into consideration, such as choice of conditional distributions to use, need for transformations of the quantities, cost and time of a simulation run and stability of the solutions obtained. These matters are also dealt with in Chapter 5.

Another scheme is given by the Metropolis-Hastings algorithms, presented in Chapter 6. They are based on a Markov chain whose dependence on the predecessor is split into two parts: a proposal and an acceptance of the proposal. The proposals suggest an arbitrary next step in the trajectory of the chain and the acceptance makes sure the appropriate limiting direction is maintained by rejecting unwanted moves of the chain. They provide a solution when even the conditional distributions of interest are complex, although their use is not restricted to these cases. Metropolis-Hastings algorithms may come in a variety of forms and these can be characterized and studied. Some of their forms may be seen as generalizations of Gibbs sampling.]

Going back to the example, the conditional distributions of α_t and β_t have also proved to be complex and Metropolis-Hastings algorithms seem to be a natural choice. Many schemes can be contemplated and a few of them are selected for numerical comparison and presented in Chapter 6.

It should also be noted that Bayesian Inference is not necessarily completed after summarizing information about unknown quantities of a given model. There may be other relevant operations to perform such as model evaluation and model comparison, involving more than one model. Of particular interest is the joint consideration of a (large) number of possible models. Bayesian Inference and MCMC can be accommodated to handle these questions. Alternative models can also be used as auxiliary devices in designing a MCMC method for a particular model. All these points are covered in Chapter 7.

Notation

Whenever possible, the same notation is maintained throughout the book. Distributions are identified with their density or probability functions and variables are generically treated as if they are continuous. Posterior densities are denoted by π and their approximations (described throughout the book) by q, observed quantities by roman letters x, y, ... and unobserved quantities or parameters by greek letters θ, ϕ, ... No distinctions are made between a random variable and its observed value and between scalar, vector and matrix quantities although matrices are generally denoted by capital letters and scalar and vector quantities by lower case letters. Vectors are always arranged in a column unless otherwise stated. The transpose of a vector x is denoted by x' and its dimension generally denoted by d.

The complement of an event A is denoted by $\bar{A}$, the probability of an event A is denoted by $Pr(A)$, and expectation and variance of a random quantity x are respectively denoted by $E(x)$ and $Var(x)$. The covariance and correlation between random quantities x and y are respectively denoted by $Cov(x, y)$ and $Cor(x, y)$. The number of elements of a set A is denoted by $\#A$. The indicator function is denoted by

$$I(x \in A) = \begin{cases} 1 & \text{, if } x \in A \\ 0 & \text{, if } x \notin A \end{cases}.$$

Approximations are denoted by a $\cdot$ superimposed to the relevant symbol. Therefore, $\doteq$ stands for approximately equal, $\overset{\cdot}{\sim}$ stands for approximately distributed as and $\overset{\cdot}{\propto}$ stands for approximately proportional to.

Components of a vector x of fixed dimension are denoted by $x_1, x_2, \ldots$ whereas elements of a sequence x will tend to be denoted by $x^{(1)}, x^{(2)}, \ldots$ This will help to distinguish between the component and the sequence dimensions when dealing with vector sequences. The identity and diagonal $d \times d$ matrices are respectively denoted by

$$I_d = \begin{pmatrix} 1 & 0 & \cdots & 0 \\ 0 & 1 & \cdots & 0 \\ \vdots & & \ddots & \vdots \\ 0 & \cdots & 0 & 1 \end{pmatrix} \text{ and } \text{diag}(c_1, \ldots, c_d) = \begin{pmatrix} c_1 & 0 & \cdots & 0 \\ 0 & c_2 & \cdots & 0 \\ \vdots & & \ddots & \vdots \\ 0 & \cdots & 0 & c_d \end{pmatrix}$$

The absolute value of the determinant of a matrix A is denoted by $|A|$.

CHAPTER 1

Stochastic simulation

1.1 Introduction

The word simulation refers to the treatment of a real problem through reproduction in an environment controlled by the experimenter. The environment is often provided by computational equipment although it can be a smaller scale reproduction of the system under study. An example is the study of tides in a laboratory reservoir. In this controlled environment, the water movement is caused by a device that is governed by an electric engine. This constitutes a study of a system that has all its components known or at least deductible and whose behaviour has an intrinsic deterministic character.

In other systems, some or all of its components are subject to random fluctuations. They cannot be described by an exact mathematical rule but only through the use of probabilistic statements. These systems are the subject matter of Statistics. Whichever way one chooses to simulate them there will be a stochastic component, namely it will be based on probability distributions. Stochastic simulation is the area of Science dealing with simulation of these systems.

From a statistical point of view, these systems may be regarded as a (possibly complicated) function of random variables. Our objective is basically to reproduce these random variables in an environment under our control regardless of the complexity of the structure relating these variables. Typically, a computer is used for this reproduction exercise.

The starting point for stochastic simulation is the construction of a random number generator. Usually, this mechanism generates an integer uniformly on an interval $[0, M]$ for a given large value of M. This generation can be reduced to a generation of a number in the unit interval $[0, 1]$ after dividing the result by M.

There are many ways of doing this. The most common is the congruential generator (Lehmer, 1951) that generates numbers according to $u_i = au_{i-1} \mod M$ after a seed u_0 is chosen. $b \mod c$ denotes the remainder of the division of b by c. In fact, there is nothing random about this sequence. One would hope that appropriate selection of the constants u_0, a and M will provide a sequence that behaves as a sequence of independent random draws from a uniform distribution on $[0, M]$. A common choice of M is $2^{31} - 1 = 2\ 147\ 483\ 647$ suggested by Lewis, Goodman and Miller (1969) and used by the IMSL library. Fishman and Moore (1985) made an exhaustive study of these generators and came up with a few suggestions for the value of a. The NAG Fortran library uses $M = 2^{59}$ and $a = 13^{13}$.

Much ingenuity and number theory has been exercised in specifying good selections of u_0, a and M. Ripley (1987) provides a nice illustrative review of the congruential generators, other number generators and the number theory results that are relevant for number generation. This text does not intend to provide a digression on the subject and the reader is referred to the books by Dagpunar (1988), Devroye (1986) and Rubinstein (1981) for further reading. Hammersley and Handscomb (1964) and Newman and Odell (1971) are useful early references on the subject.

From now on, it will be assumed that the reader has a random number generator on the unit interval available to him. The quantity generated will be denoted by u and will be the base for all future random generations. The words generated, drawn or even sampled are used indistinguishably in this book to describe a generation (or draw or sample) from a distribution.

It must be stressed that all methods currently available are based on iterative and deterministic mechanisms. It is more accurate to refer to them as pseudo-random to clarify that these mechanism can only generate numbers with a similar behaviour to truly random numbers. However, if the mechanism is not known and examined only through its output, it provides numbers with the important basic properties of random numbers: uniformity and independence.

The next section describes techniques that can be used for the generation of some discrete distributions. The same is done in section 1.3 for continuous distributions. Section 1.4 then deals with multivariate distributions. In section 1.5, techniques based on a auxiliary generation from an approximate distribution are described. These techniques are particularly useful if the target distribution has a complicated form and there are no direct methods of generation of the random variable or quantity[*].

1.2 Generation of discrete random quantities

This section describes methods of generation of quantities with discrete probability distribution. This is based on a random quantity u generated from a uniform distribution on the unit interval $[0, 1]$, denoted by $u \sim U[0, 1]$.

In the generic case of a quantity x assuming values on $\{x_1, ..., x_k\}$ with respective probabilities $p_1, ..., p_k$ subject to $\Sigma_i p_i = 1$, the interval $[0, 1]$ is split into k intervals $I_1, ... I_k$ with $I_i = (F_{i-1}, F_i]$ where $F_0 = 0$ and

---

[*] The terms quantity and variable are used indistinctly as in the context of this book it is possible to specify probability distributions to observable variables and fixed but unobservable quantities. This point is clarified in the next chapter, but to emphasize it the word quantity will be used whenever possible.

$F_i = p_1 + ... + p_i$, $i = 1, ..., k$. Each interval corresponds to a single value for x and after observing the generated value of u, one verifies the interval I_i to which u belongs. The generated value for x is x_i. The method generates values from the distribution of x because $Pr(x = x_i) = Pr(u \in I_i) = F_i - F_{i-1} = p_i$, for $i = 1, ..., k$. In the case of a discrete uniform distribution, it is enough to divide the unit interval into k subintervals of equal length $1/k$ and $F_i = i/k$. This construction can also be used for distributions taking values on a countable set by extending the argument.

In both the finite for large k and countable case, there is an important computational problem. It relates to the number of comparisons required to find the appropriate interval. Many comparisons may be needed before the correct interval is chosen. One possible search scheme to speed computations is to reorder the possible values of x in decreasing order of their probabilities. Most likely values will require less comparisons. Binary and indexed searches may also be used (Ripley, 1987).

For many distributions there is a structure relating the values of the probabilities p_i to a smaller number of constants. It is possible to obtain more efficient schemes without evaluation of the cumulative probabilities F_i in these cases. Below, some of these methods are described for the better known distributions.

a) Bernoulli distribution

x has Bernoulli distribution with success probability p if $Pr(x = 0) = 1 - p$ and $Pr(x = 1) = p$, $0 \leq p \leq 1$, denoted by $x \sim bern(p)$. In this case, the general result above can be used with $k = 2$, $x_1 = 1$, $x_2 = 0$ and $F_1 = p$. In other words, the unit interval is divided into two pieces of lengths p and $1 - p$. If the value of u is smaller than or equal to p, the value 1 is returned for x. Otherwise, the value 0 is returned.

b) Binomial distribution

x has binomial distribution with index n and success probability p, denoted by $bin(n, p)$, if it has probability function

$$f_B(i) = Pr(x = i) = \binom{n}{i} p^i (1 - p)^{n-i}, \quad i = 0, 1, ..., n$$

The generic method can be used here but would involve tedious calculations of combinations which may involve overflow difficulties in some computers if n is large. A more efficient and easier to code solution is obtained by noting that if $x_1, ..., x_n$ form a sample from the $bern(p)$ distribution, that is, are independently and identically distributed, then $x = \Sigma_i x_i \sim bin(n, p)$. So, a value x drawn from a binomial distribution is obtained by drawing a sample $u_1, ..., u_n$ from a $U[0, 1]$ distribution and counting the number x of these n generated values that are smaller than or equal to p. The resulting value has $bin(n, p)$ distribution.

c) Geometric and negative binomial distribution

x has negative binomial distribution with index r and success probability

p, denoted by $nb(r, p)$, if it has probability function

$$f_{NB}(n) = Pr(x = n) = \binom{n-1}{r-1} p^r (1-p)^{n-r}, \quad n = r, r+1, r+2, \ldots$$

If $r = 1$, the distribution is called geometric and denoted $geom(p)$.

As in (b), relations with the Bernoulli distribution may be invoked to simplify the problem. It is easy to see that a $nb(r, p)$ quantity is obtained by repeating $bern(p)$ observations until r successes are obtained and counting the number of observations used. So, a $nb(r, p)$ quantity is drawn through the generation of a sequence $u_1, u_2, \ldots$ from $U[0, 1]$ until r of them are smaller than or equal to p and counting the number x of generated quantities u_i.

d) Poisson distribution

x has Poisson distribution with mean λ, denoted by $Poi(\lambda)$, if it has probability function

$$f_P(i) = Pr(x = i) = e^{-\lambda} \frac{\lambda^i}{i!}, \quad i = 0, 1, 2, \ldots$$

The most efficient method of generation of $Poi(\lambda)$ quantities with small to moderate values of λ is based on generation of exponential random quantities. x has exponential distribution with parameter λ, denoted by $Exp(\lambda)$, if it has density $f(x) = \lambda e^{-\lambda x}$, $x > 0$. The generation of exponential quantities will be presented in the next section as this is a continuous distribution. For the moment, simply assume that a sample $y_1, y_2, \ldots$ from the $Exp(\lambda)$ can be drawn.

Results relating the Poisson and exponential distributions are obtained in Poisson processes. These are counting processes $\{N(t), t \geq 0\}$ where $N(t)$ counts the number of occurrences of a given event from the origin up to time t. The Poisson process is defined through:

(1). $N(0) = 0$;

(2). $N(s)$ and $N(t) - N(s)$ are independent quantities, for $0 < s < t$;

(3). $N(t) \sim Poi(\lambda t)$.

Denote the times between successive occurrences by $y_1, y_2, \ldots$ and the times of the occurrences by $t_1 = y_1, t_2 = y_1 + y_2, \ldots$. By (2), $N(t_1), N(t_2) - N(t_1), N(t_3) - N(t_2), \ldots$ are independent quantities. Hence, $y_1, y_2, \ldots$ are also independent quantities. For $t > 0$, $Pr(y_i > t) = Pr(N(t_{i-1} + t) - N(t_{i-1}) = 0) = e^{-\lambda t}$ as, by (3), $N(t_{i-1} + t) - N(t_{i-1}) \sim Poi(\lambda t)$. Therefore, $F(t) = Pr(y_i \leq t) = 1 - e^{-\lambda t}$ and the common density of the y_i is $f(t) = \lambda e^{-\lambda t}$, $t > 0$. Times between occurrences of a Poisson process form a sample from the $Exp(\lambda)$ and $N(1) \sim Poi(\lambda)$.

The generation of $Poi(\lambda)$ random quantities is based on generations of $Exp(\lambda)$ until the sum of these quantities, playing the role of interoccurrences times, goes beyond 1. If the number of generations required to this end is $k + 1$, the number of occurrences $N(1)$ and consequently the value of x drawn is k.

1.3 Generation of continuous random quantities

1.3.1 Probability integral transform

The basic result about generation of continuous quantities is the probability integral transform stating that if x is continuous with distribution function F then $u = F(x) \sim U[0, 1]$. This result is easily shown as

$$F_u(y) = Pr(u \leq y) = Pr(F(x) \leq y) = Pr(x \leq F^{-1}(y)) = F(F^{-1}(y)) = y,$$

if $0 < y < 1$. Exploring the fact that F has an inverse, one finds that $F^{-1}(u) \sim F$. So, a quantity with distribution function F is drawn after transforming a uniform quantity according to F^{-1}. Below, some applications of this method are presented.

a) Exponential distribution

This distribution was introduced in the previous section. All that is left here is to obtain the expression of F^{-1}. As $F(x) = 1 - e^{-\lambda x}$,

$$\begin{aligned}
u &= 1 - e^{-\lambda x} \Leftrightarrow \\
1 - u &= e^{-\lambda x} \Leftrightarrow \\
\log(1 - u) &= -\lambda x \Leftrightarrow \\
x &= -\tfrac{1}{\lambda} \log(1 - u)
\end{aligned}$$

Generation of x may be simplified further by noting that $1 - u \sim u$ and therefore $-(1/\lambda) \log u$ leads to a value from $Exp(\lambda)$.

b) Weibull distribution

The Weibull distribution can be obtained as a generalization of the exponential distribution and is frequently used in reliability studies. If $x \sim Exp(\lambda)$ then $y = x^{1/\alpha}$ has Weibull distribution with parameters α and λ. Its generation is therefore trivial requiring simply a draw u from a uniform distribution and returning a value $[-(1/\lambda) \log u]^{1/\alpha}$.

c) Gamma distribution

x has Gamma distribution with parameters α and β, denoted by $G(\alpha, \beta)$, if its density is

$$f_G(x; \alpha, \beta) = \frac{\beta^\alpha}{\Gamma(\alpha)} x^{\alpha-1} e^{-\beta x}, \quad x > 0$$

If $x_1, ..., x_n$ is a sample from a $Exp(\lambda)$ distribution then $x = x_1 + ... + x_n \sim G(n, \lambda)$ (DeGroot, 1986, p. 289). So, a quantity having Gamma distribution with $\alpha = n$ integer can be generated by drawing a sample of size n from the exponential distribution and summing up the sample values.

The χ^2 distribution with n degrees of freedom, denoted by χ_n^2, is given by the $G(n/2, 1/2)$ distribution. Furthermore it is related to the normal distribution by the following result: if $x_1, ..., x_n$ form a sample from the standard normal distribution then $x = x_1^2 + \ldots + x_n^2 \sim \chi_n^2$ (Mood, Gray-bill and Boes, 1974, p. 242). So, generation of a χ_n^2 distribution involves generation of a standard normal sample of size n and summation of their squares. This scheme depends on generation of normal quantities, to be presented in the next section. It extends the above generation of Gamma distributions by including those with α being an integer multiple of $1/2$. Generation of Gamma quantities with an arbitrary value of α is deferred to section 1.5.

The parameter β of the Gamma distribution is a scale parameter which implies that if $x_1 \sim G(\alpha, \beta_1)$ and $x_2 \sim G(\alpha, \beta_2)$ then $\beta_1 x_1$ and $\beta_2 x_2$ have the same $G(\alpha, 1)$ distribution. This is a special case of a general result valid for any scale model (Exercise 1.5). A quantity $y \sim G(\alpha, \beta)$ can be generated from $x \sim G(\alpha, 1)$ by taking $y = x/\beta$.

If $x \sim G(\alpha, \beta)$ then x^{-1} has inverted Gamma distribution with parameters α and β, denoted by $IG(\alpha, \beta)$. Generation of an IG quantity simply involves generation of a Gamma quantity with the same parameters and taking its inverse. The difficulty is therefore the same as that with a Gamma generation.

1.3.2 Bivariate techniques

If (x_1, x_2) has joint density $f_x(x_1, x_2)$ and $g(x_1, x_2) = (y_1, y_2)$ is a one-to-one differentiable transformation with inverse $g^{-1}(y_1, y_2) = (x_1, x_2)$, then the density of (y_1, y_2) is

$$f_y(y_1, y_2) = f_x(g^{-1}(y_1, y_2)) \, J \text{ where } J = \begin{vmatrix} \partial x_1/\partial y_1 & \partial x_2/\partial y_1 \\ \partial x_1/\partial y_2 & \partial x_2/\partial y_2 \end{vmatrix}$$

Although the result concerns bivariate vectors, it is often useful for generation of scalar quantities.

1) Normal distribution

x has normal distribution with mean μ and variance σ^2, denoted by $N(\mu, \sigma^2)$, if its density is

$$f_N(x; \mu, \sigma^2) = \frac{1}{\sqrt{2\pi\sigma^2}} \exp\left\{-\frac{1}{2\sigma^2}(x - \mu)^2\right\}$$

The standard normal distribution is obtained when $\mu = 0$ and $\sigma^2 = 1$.

As will be seen in the next chapter, it is particularly advantageous in the Bayesian context to work with the reparametrization $\phi = 1/\sigma^2$. The parameter ϕ, the inverse of the variance, is usually referred to as precision. For this parametrization, the density is

$$f_N(x; \mu, \phi^{-1}) = \frac{\phi^{1/2}}{\sqrt{2\pi}} \exp\left\{-\frac{\phi}{2}(x - \mu)^2\right\}$$

The normal distribution is possibly the most used distribution in statistical applications. It is therefore fundamental to have a fast method of generation of normal quantities. Unfortunately, all methods available require at least two uniform quantities.

Box and Muller (1958) showed that if u_1 and u_2 are independent $U[0,1]$ quantities, $x_1 = \sqrt{-2\log u_1}\cos(2\pi u_2)$ and $x_2 = \sqrt{-2\log u_1}\sin(2\pi u_2)$ then x_1 and x_2 are independent $N(0,1)$ quantities. The result is proved by constructing a transformation

$$(x_1, x_2) = g(u_1, u_2) = (\sqrt{-2\log u_1}\cos(2\pi u_2), \sqrt{-2\log u_1}\sin(2\pi u_2)).$$

This is clearly a one-to-one transformation taking points from the unit square $[0,1]^2$ into R^2 with inverse $u_1 = \exp\{-(x_1^2 + x_2^2)/2\}$ and $u_2 = (1/2\pi)\tan^{-1}(x_2/x_1)$. As $f_u(u_1, u_2) = 1$ for $(u_1, u_2) \in [0,1]^2$, and 0 otherwise, $f_u(g^{-1}(x_1, x_2)) = 1$ for $(x_1, x_2) \in R^2$. Therefore, $f_x(x_1, x_2) = J$ for $(x_1, x_2) \in R^2$ and

$$
\begin{aligned}
J &= \left| \begin{array}{cc} x_1 \exp\{-(x_1^2 + x_2^2)/2\} & (1/2\pi)\{1/[1 + (x_2/x_1)^2]\}(-x_2/x_1^2) \\ x_2 \exp\{-(x_1^2 + x_2^2)/2\} & (1/2\pi)\{1/[1 + (x_2/x_1)^2]\}(1/x_1) \end{array} \right| \\
&= \left| \begin{array}{cc} x_1 \exp\{-(x_1^2 + x_2^2)/2\} & -(1/2\pi)[x_2/(x_1^2 + x_2^2)] \\ x_2 \exp\{-(x_1^2 + x_2^2)/2\} & (1/2\pi)[x_1/(x_1^2 + x_2^2)] \end{array} \right| \\
&= \frac{1}{2\pi}\exp\left\{ -\frac{x_1^2 + x_2^2}{2} \right\}
\end{aligned}
$$

So, $f_x(x_1, x_2) = f(x_1)f(x_2)$ where $f(x_i) = (1/\sqrt{2\pi})\exp\{-x_i^2/2\}$, $i = 1, 2$, proving the result. Another method based on pairs of uniform quantities is left as an exercise.

Finally, it is worth mentioning that a more natural and simpler method of normal generation is based on the central limit theorem. Consider a sample $u_1, ..., u_n$ from the $U[0,1]$ distribution. Then, for large enough n,

$$x = \sqrt{n}\,\frac{\bar{u} - 1/2}{1/\sqrt{12}} \,\dot\sim\, N(0,1) \text{ where } \bar{u} = \frac{1}{n}\sum_{i=1}^{n} u_i$$

A natural choice of n is 12 to simplify the expression of x but this is generally too small to yield a reasonable approximation, especially in the tails. Despite the historical interest, this method is very inefficient, requiring much more computing time than the Box-Muller method.

2) Ratio-of-uniforms method

This is a very general method proposed by Kinderman and Monahan (1977) for generation of quantities with an arbitrary density f that may be known up to a proportionality constant. Hence, the method generates values of a density from which only the algebraic kernel is known. Let x_1 and x_2 be quantities uniformly distributed in the region $C_f = \{(x_1, x_2) : 0 \geq x_1 \geq \sqrt{f^*(x_2/x_1)}\}$. Then, $y = x_2/x_1$ has a distribution with density $f(y) = f^*(y)/\int f^*(u)du$.

Again, the result can be obtained by completing the transformation with

$z = x_1$. The inverse is given by $x_1 = z$ and $x_2 = yz$ with $|J| = z$. As $f(x_1, x_2) = k$ for $(x_1, x_2) \in C_f$ where $k^{-1} = \text{area}(C_f)$, then $f(y, z) = kz$, $0 \geq z \geq \sqrt{f^*(y)}$. Integrating with respect to z,

$$f(y) = \int_0^{\sqrt{f^*(y)}} kz \, dz = \frac{k}{2} z^2 \big|_0^{\sqrt{f^*(y)}} = \frac{k}{2} f^*(y)$$

Therefore, y has density proportional to f^*. The uniform generation from C_f may be difficult for complicated forms of f. This may be simplified (by graphical techniques) whenever it is possible to find the constants $a = \sqrt{\sup_x f^*(x)} > 0$, $b_1 = \sqrt{\inf_{x \leq 0} x^2 f^*(x)}$ and $b_2 = \sqrt{\sup_{x \geq 0} x^2 f^*(x)}$ such that $C_f \subset [0, a] \times [b_1, b_2]$. Generation of a pair (y_1, y_2) of independent uniform quantities on $[0, a] \times [b_1, b_2]$ is easy by linear transformation. Passage from a uniform distribution on $[0, a] \times [b_1, b_2]$ to a uniform distribution on C_f is achieved by rejection methods (see section 1.5).

A very simple application of this method is the generation from the Cauchy distribution (see definition at the end of this section) by uniform generation of points in the unit semi-circle. Other important applications are the generation of Gamma (see also section 1.5 and Exercise 1.14) and normal variates.

Wakefield, Gelfand and Smith (1991) studied generalizations of the method. They first considered generation of a pair (x_1, x_2) uniformly distributed in the region $C_{f,g} = \{(x_1, x_2) : 0 \geq x_1 \geq g^{-1}[cf^*(x_2/g'(x_1))]\}$ with a strictly increasing differentiable function g and a constant $c > 0$ and showed that $y = x_2/g'(x_1)$ has a distribution with density $f(y) = f^*(y)/\int f^*(u)du$. A popular choice for g is the power family $g(x) = x^{r+1}/(r+1)$ for $r \geq 0$ and $c = 1/(r+1)$. If $r = 1$, the original ratio-of-uniforms method is obtained and $r = 0$ corresponds to a uniform envelope studied in Example 1.2 below. They recommend using $r = 0.5$ which is optimal for the normal case in the sense of minimizing expected generation cost (Exercise 1.9) and behaves well in some other applications. They also show that relocation of f^* to have mode at zero is optimal. A multivariate generalization and proof of correctness of the algorithms are also left as exercises.

3) Logarithmic distribution

Finally, another interesting, specific application of bivariate techniques is presented. x has logarithmic distribution if its density is given by $f(x) = -\log x$ for $0 < x < 1$, and 0 otherwise. The probability integral transform does not have analytic solution but if u_1 and u_2 are independent $U[0, 1]$ quantities then $x = u_1 u_2$ has logarithmic distribution. To prove the result, complete the transformation g with $y = u_1$. The inverse g^{-1} is given by $(y, x/y)$ with jacobian

$$J = \begin{vmatrix} 0 & 1/y \\ 1 & -x/y^2 \end{vmatrix} = 1/y$$

So, $f(x, y) = 1/y$, for $0 < x < y < 1$, and 0 otherwise. Then,

$$f(x) = \int_x^1 \frac{1}{y} dy = \log y \,|_x^1 = -\log x \quad, \quad 0 < x < 1$$

1.3.3 Methods based on mixtures

Methods based on mixtures are also bivariate techniques. The motivation behind the method, however, is different with the original variables playing different roles. In essence, methods based on mixtures explore the simple fact that the joint density of (x_1, x_2) satisfies

$$f(x_1, x_2) = f(x_1|x_2)f(x_2)$$

Therefore, the pair (x_1, x_2) can be generated in two steps: first, x_2 is generated and then x_1 is generated conditionally on the generated value of x_2. This mechanism automatically provides a value x_1 from its marginal distribution. Despite its simplicity, this result is of great importance in simulation. By breaking down the joint distribution, one is able to generate a value for x_1 even when its marginal density is unavailable analytically or at least awkward to sample from directly.

The quantity x_2 is the mixing element. When it is discrete with probability function $p(a_i) = p_i$, $\forall i$ then the marginal density of x_1 is given by

$$f(x_1) = \sum_i p_i f_i(x_1) \text{ where } f_i(x_1) = f(x_1|x_2 = a_i)$$

If all f_i are equal, then they are also equal to f irrespective of the mixture weights p_i. A common choice in this case is $p_i = 1/k$ where k is the number of components in the mixture. This mixing scheme provides another form of generation from f that improves the randomness of the pseudorandom generators without altering the target distribution.

A context where the occurrence of mixture is commonplace is in robustness studies. Instead of assuming the usual $N(\mu, \sigma^2)$ model as appropriate, one assumes that the model is appropriate with probability $1 - \epsilon$, for small value of ϵ. There is also a probability ϵ that the appropriate model is *corrupted* by a shift in level $N(\mu + a, \sigma^2)$ or by an inflated variance $N(\mu, k\sigma^2)$, $k > 1$. In the first case,

$$f(x_1) = (1 - \epsilon)f_N(x_1; \mu, \sigma^2) + \epsilon f_N(x_1; \mu + a, \sigma^2)$$

A discrete indicator variable x_2 is constructed to indicate the component of the mixture to be chosen. If $x_2 = 0$, x_1 is drawn from the standard component but if $x_2 = 1$, x_1 is drawn from the corrupted component. The model is completed by the specification of a *bern*(ϵ) distribution for x_2. Similar calculations can be made for the case of variance contamination.

Discrete mixtures are also used in the fragmentation of a generation in many component parts, each having a given probability of being chosen. This technique is useful when some components are easy to generate from and concentrate most of the probability. In this case, a costly generation is replaced by another one that is cheap with high probability and only remains costly for low probability components.

Assume that one wishes to draw a value from a density f but one knows how to sample from a density f_1 satisfying $f(x) - a_1 f_1(x) \geq 0$ for some $a_1 > 0$. It follows that

$$\int (f(x) - a_1 f_1(x))\,dx = 1 - a_1$$

and therefore, a density $g_1(x) = (f(x) - a_1 f_1(x))/(1 - a_1)$ may be defined. This implies a mixture expression $f(x) = a_1 f_1(x) + (1 - a_1)g_1(x)$. As draws from f_1 are easy to obtain, it is advantageous that a_1 be as large as possible subject to the requirements that $a_1 \leq 1$ and $f - a_1 f_1 \geq 0$. This implies a value for a_1 given by the minimization of $f(x)/f_1(x)$.

All that is left is the problem of generation from g_1. Here again the same process of decomposition can be applied by choosing a density f_2 easy to draw from and satisfying $g_1(x) - b_2 f_2(x) \geq 0$. Again, it is advantageous to take the largest possible value for b_2 (given by $\min_x g_1(x)/f_2(x)$) and to create $g_2(x) = (g_1(x) - b_2 f_2(x))/(1 - b_2)$ such that $g_1(x) = b_2 f_2(x) + (1 - b_2)g_2(x)$. Replacing g_1 in the expression of f,

$$
\begin{aligned}
f(x) &= a_1 f_1(x) + (1 - a_1)b_2 f_2(x) + (1 - a_1)(1 - b_2)g_2(x) \\
     &= a_1 f_1(x) + a_2 f_2(x) + (1 - a_1 - a_2)g_2(x)
\end{aligned}
$$

where $a_2 = (1 - a_1)b_2$. This process of decomposition of densities may proceed indefinitely with succesive choices of approximating densities f_{i+1} for g_i, $i = 2, 3, \ldots$ with weights b_{i+1}. The corresponding weights of f_{i+1} in the expression for f are given by $a_{i+1} = (1 - a_1 - \ldots - a_i)b_{i+1}$, $i = 1, 2, \ldots$ and a_1 for f_1. If the weights $a_1, a_2, \ldots$ are well chosen, the residual densities g_i, possibly more costly, will have negligible effect on the mixture and will not affect the efficiency of the method.

Example 1.1 This technique was applied by Marsaglia and Bray (1964) to the generation from the $N(0, 1)$ distribution. Let $x_1 = 2(u_1 + u_2 + u_3 - 1.5)$ where the u_i are independent $U(0,1)$. Note that $E(x_1) = 0$ and $Var(x_1) = 1/3$ but $|x_1| \leq 3$. The density of x_1 is

$$f_1(x) = \begin{cases} (3 - x^2)/8, & |x| \leq 1 \\ (3 - |x|)^2/16, & 1 < |x| \leq 3 \end{cases}$$

and 0 otherwise. The largest possible value for a_1 is given by $\min_x f(x)/f_1(x)$ $= f(2)/f_1(2) = 0.8638$. So, x_1 may be used to generate $N(0, 1)$ variates more than 86% of the time. The residual g_1 may be approximated by the triangular density

$$f_2(x) = (6 - 4|x|)/9, \quad |x| \le 1.5$$

and 0 otherwise, which is the density of $x_2 = 1.5(u_4 + u_5 - 1)$ where the u_i are also independent $U(0,1)$. The largest possible value for b_2 is given by $min_x g_1(x)/f_2(x) = g_1(0.8739)/f_1(0.8739) = 0.8139$ and $a_2 = (1 - a_1)b_2 = 0.1107$. So, the components x_1 and x_2 that are very easy to obtain are responsible for generation at $86.38\% + 11.07\% = 97.45\%$ of the time. The remaining probability is split into two parts: the remaining densities f_3, defined for $|x| > 3$, and f_4 in the interval $[-3, 3]$. The density f_3 is given by the $N(0, 1)$ density restricted to $|x| > 3$ which has probability $a_3 = \int_{|x|>3} f_N(x; 0, 1)dx = 0.0027$. The density f_4 is given by

$$f_4(x) = \frac{f_n(x; 0, 1) - a_1 f_1(x) - a_2 f_2(x) - a_3 f_3(x)}{1 - a_1 - a_2 - a_3}$$

with weight $a_4 = 1 - a_1 - a_2 - a_3 = 0.0228$. Generation from f_3 and f_4 is acomplished via rejection techniques. Details of these generation procedures are left as an exercise.

If the quantity x_2 is continuous, the marginal density of x_1 is

$$f(x_1) = \int f(x_1|x_2)f(x_2)dx_2$$

An example that appears many times in this book is a scale mixture of normal densities (Andrews and Mallows, 1974) where $f(x_1|x_2) = N(\mu, x_2\sigma^2)$. Many distributions may be obtained in this way. The most famous one is the Student's t distribution but the double exponential, logistic and stable distributions can also be obtained as scale mixtures.

x has Student's t distribution with n degrees of freedom $(n > 0)$, mean μ $(\mu \in R)$ and scale parameter σ^2 $(\sigma^2 > 0)$, denoted by $t_n(\mu, \sigma^2)$, if its density is

$$f_t(x; n, \mu, \sigma^2) = \frac{\Gamma[(n+1)/2]}{\Gamma(n/2)\Gamma(1/2)} \frac{1}{\sqrt{n\sigma^2}} \left[1 + \frac{1}{n}\left(\frac{x - \mu}{\sigma}\right)^2\right]^{-(n+1)/2}$$

When $n = 1$, the distribution is called the Cauchy distribution.

The joint distribution of a random pair (x_1, x_2) is called normal-Gamma with parameters μ, σ^2, n and S, denoted by $NG(\mu, \sigma^2, n, S)$, if $x_1|x_2 \sim N(\mu, \sigma^2/x_2)$ and $x_2 \sim G(n/2, nS/2)$. An alternative representation is $x_1|x_2 \sim N(\mu, \sigma^2 x_2)$ and $x_2 \sim IG(n/2, nS/2)$. It can be shown that the distribution of x_1 is $t_n(\mu, S\sigma^2)$ for both representations (see, for example, DeGroot, 1970). So, generation of a $t_n(\mu, \sigma^2)$ can be made in two steps: first, the scale x_2 is drawn from a $G(n/2, n/2)$ (taking $S = 1$) and then x_1 is drawn from a $N(\mu, \sigma^2/x_2)$ with the value of x_2 generated in the first step.

1.4 Generation of random vectors and matrices

As for univariate quantities, the main generation techniques for random vectors and matrices are the probability integral transform and general transformations in the style of those in the previous section. The first class of transformation is used to generate independent univariate quantities on which specific transformations are operated to introduce correlation between the elements of a random vector or matrix.

If a vector $x = (x_1, ..., x_d)'$ has joint density $f_x(x_1, ..., x_d)$ and $g(x_1, ..., x_d) = (y_1, ..., y_d)'$ is a differentiable one-to-one transformation with inverse $g^{-1}(y_1, ..., y_d) = (x_1, ..., x_d)'$, the density of $(y_1, ..., y_d)'$ is

$$f_y(y_1, ..., y_d) = f_x(g^{-1}(y_1, ..., y_d))\, J$$

where

$$J = \begin{vmatrix} \partial x_1/\partial y_1 & \cdots & \partial x_d/\partial y_1 \\ \vdots & & \vdots \\ \partial x_1/\partial y_n & \cdots & \partial x_d/\partial y_d \end{vmatrix}$$

1) Multivariate normal distribution

$x = (x_1, ..., x_d)'$ has multivariate (or d-variate) normal distribution with mean μ and variance Σ, denoted by $N(\mu, \Sigma)$ or $N_d(\mu, \Sigma)$ when specification of the dimension is useful, if its density is

$$f_N(x; \mu, \Sigma) = (2\pi)^{p/2} |\Sigma|^{1/2} \exp\left\{ -\frac{1}{2}(x-\mu)'\Sigma^{-1}(x-\mu) \right\}.$$

The variance matrix Σ must have full rank d for the above density to exist. Nevertheless, there are some advantages in keeping the notation even when Σ is not of full rank. The standard normal distribution is obtained when $\mu = 0$ and $\Sigma = I_d$, the identity matrix of order d. In this case, it is easy to see that the components x_i are independent standard normal quantities. The univariate normal distribution is the special case where $d = 1$.

The normal distribution possesses many properties that are particularly attractive from an applied point of view, making it one of the main probability distributions. The most important properties for the development of this text are

Linear transformations: if A is a $r \times d$ matrix of constants and b is a r-dimensional vector of constants and $x \sim N_d(\mu, \Sigma)$ then

$$y = Ax + b \sim N_r(A\mu + b, A\Sigma A') \tag{1.1}$$

Marginal distributions: if the vector x is divided into 2 blocks, x_1 containing the first d_1 components of x and x_2 containing the other $d_2 = d - d_1$ components, then applying a similar partition on μ and Σ in the form

$$\mu = \begin{pmatrix} \mu_1 \\ \mu_2 \end{pmatrix} \text{ and } \Sigma = \begin{pmatrix} \Sigma_{11} & \Sigma_{12} \\ \Sigma_{21} & \Sigma_{22} \end{pmatrix}$$

leads to $x_i \sim N_{d_i}(\mu_i, \Sigma_{ii})$, $i = 1, 2$.

Conditional distributions: still using the same partitions on x, μ and Σ

$$x_1|x_2 \sim N_{d_1}(\mu_{1\cdot 2}, \Sigma_{11\cdot 2}) \tag{1.2}$$

where $\mu_{1\cdot 2} = \mu_1 + \Sigma_{12}\Sigma_{22}^{-1}(x_2 - \mu_2)$ and $\Sigma_{11\cdot 2} = \Sigma_{11} - \Sigma_{12}\Sigma_{22}^{-1}\Sigma_{21}$. Analogous results are obtained for the distribution of $x_2|x_1$ by exchanging all indices with values 1 and 2. For these results, it is obviously required that the submatrices Σ_{22} and Σ_{11} respectively are of full rank, otherwise their inverses will not exist.

Reconstruction of the joint distribution: if $x_1|x_2 \sim N_{d_1}(\mu_1 + B_1(x_2 - \mu_2), B_2)$ for $d_1 \times d_2$ and $d_1 \times d_1$ matrices of constants B_1 and B_2 respectively and $x_2 \sim N_{d_2}(\mu_2, \Sigma_{22})$ then

$$x = \begin{pmatrix} x_1 \\ x_2 \end{pmatrix} \sim N_d \left[\begin{pmatrix} \mu_1 \\ \mu_2 \end{pmatrix}, \begin{pmatrix} \Sigma_{11} & \Sigma_{12} \\ \Sigma_{21} & \Sigma_{22} \end{pmatrix} \right] \tag{1.3}$$

where $\Sigma_{11} = B_2 + B_1\Sigma_{22}B_1'$ and $\Sigma_{21}' = \Sigma_{12} = B_1\Sigma_{22}$.

Quadratic forms: $(x - \mu)'\Sigma^{-1}(x - \mu) \sim \chi_d^2$.

These results can be found in standard multivariate analysis textbooks such as Anderson (1958).

Generation of a d-variate normal random vector $y \sim N(\mu, \Sigma)$ uses the techniques described at the beginning of this section. Initially, d independent $N(0, 1)$ random quantities are generated according to one of the techniques described in the previous section thus forming a d-dimensional vector x with a $N(0, I_d)$ distribution. A $N(\mu, \Sigma)$ variate is obtained after using the transformation (1.1) with $b = \mu$ and $\Sigma = AA'$. The $d \times d$ matrix A is the square root matrix of Σ and there are many algorithms for finding a solution for A. One of them is the singular value decomposition. It prescribes how to obtain an orthogonal matrix Q whose columns are the eigenvectors of Σ and a diagonal matrix Λ with main diagonal containing the eigenvalues of Σ such that $\Sigma = Q\Lambda Q' = PP'$ where $P = Q\Lambda^{1/2}$. Another algorithm with possibly lower computational cost is the Choleski decomposition that finds the unique solution for A in the convenient form of a lower triangular matrix (see Chambers (1977) for details of the calculations).

2) Wishart distribution

A $d \times d$ symmetric and positive-definite matrix X has Wishart distribution with α degrees of freedom and parameter β, where $\alpha > (d-1)/2$ and β is a $d \times d$ symmetric and positive-definite matrix, denoted by $W(\alpha, \beta)$ or $W_d(\alpha, \beta)$ when specification of the dimension is useful, if its density is

$$f_W(X; \alpha, \beta) = \frac{|\beta|^\alpha}{\pi^{d(d-1)/4} \prod_{i=1}^d \Gamma\left(\alpha - \frac{i-1}{2}\right)} |X|^{\alpha - (d+1)/2} \exp\{-tr(\beta X)\}$$

for positive-definite values of X and 0 otherwise. Note that if $d = 1$ and therefore X is scalar, the density above reduces to the $G(\alpha, \beta)$ density.

It can be shown that if $x_1, ..., x_n$ are a sample from a $N_d(\mu, \Sigma)$ then

$$\sum_{i=1}^{n}(x_i - \mu)(x_i - \mu)' \sim W_d(n/2, \Sigma^{-1}/2)$$

So, an easy generation method for the $W_d(\alpha, \beta)$ is available from a sample of (integer) size $n = 2\alpha$ from the $N_d(\mu, \beta^{-1}/2)$ distribution. As in the univariate case of Gamma generations, this normal based method only allows generation of Wishart matrices with integer multiples of $1/2$ degrees of freedom.

Defining now $y_i = A(x_i - \mu)$, $i = 1, ..., n$ where A is the square root matrix of Σ implies that $y_1, ..., y_n$ are a sample from the $N_d(0, I_d)$ and $\Sigma_i y_i y_i' \sim W_d(n/2, I_d/2)$. On the other hand,

$$\sum_{i=1}^{n} y_i y_i' = \sum_{i=1}^{n}[A(x_i - \mu)][A(x_i - \mu)]' = A\sum_{i=1}^{n}(x_i - \mu)(x_i - \mu)'A'$$

So, if $x \sim W_d(\alpha, \beta)$ then $AxA' \sim W_d(\alpha, A\beta A') = W_d(\alpha, I_d)$ if $\beta = AA'$. The condition is also suffucent in that if $y \sim W_d(\alpha, I_d)$ and β is a positive-definite matrix and admits the decomposition $\beta = AA'$ then $AyA' \sim W_d(\alpha, \beta)$.

This result allows generation of Wishart matrices from independent normal variates. The basic form to draw a $d \times d$ matrix $x \sim W(n/2, I_d/2)$ is to generate independent dn $N(0, 1)$ quantities $x_{11}, ..., x_{1n}, ..., x_{d1}, ..., x_{dn}$, form a sample of vectors $x_i = (x_{1i}, ..., x_{di})' \sim N_d(0, I_d)$, $i = 1, ..., n$ and

$$Y = \sum_{i=1}^{n} x_i x_i' \sim W_d(n/2, I_d/2)$$

If a matrix $X = (x_1, ..., x_n)$ is constructed then $Y = XX'$ and more efficient methods using other decompositions and possibly using a smaller number of variate generations may be devised.

Odell and Feiveson (1966) explored this idea using the decomposition $Y = LL'$ where L is a $d \times d$ lower triangular matrix. They showed that if the main diagonal elements l_{ii} of L are square roots of $G((n-i+1)/2, 1/2)$ (or χ^2_{n-i+1}) quantities, the off-diagonal elements l_{ij} of L are $N(0, 1)$ quantities and if all these variates are independently drawn then $Y \sim W_d(n/2, I_d/2)$. This method only involves $d(d - 1)/2$ normal draws and d Gamma draws which implies great computational savings when compared to the *brute force* method that requires dn normal draws. The larger the values of d and n, the greater the computational savings. Another side advantage of this method is to allow easy calculation of the determinant of Y since $|Y| = |LL'| = (|L|)^2 = (\prod_i l_{ii})^2$.

If X has $W(\alpha, \beta)$ distribution then X^{-1} is said to have inverted Wishart distribution with the same parameters, denoted by $IW(\alpha, \beta)$ or $IW_d(\alpha, \beta)$. As in the univariate case with the Gamma distribution, a random matrix with $IW(\alpha, \beta)$ distribution is obtained by drawing a $W(\alpha, \beta)$ matrix and inverting it.

3) Multivariate Student's t distribution

x has multivariate (or d-variate) Student's t distribution with n degrees of freedom, mean μ and scale parameter Σ where $n > 0$, $\mu \in R^d$ and Σ is a $d \times d$ symmetric and positive-definite matrix, denoted by $t_n(\mu, \Sigma)$, if its density is

$$f_t(x; n, \mu, \Sigma) = \frac{\Gamma[(n+d)/2]}{\Gamma(n/2)(n\pi)^{d/2}} |\Sigma|^{-1/2} \left[1 + \frac{1}{n}(x-\mu)'\Sigma^{-1}(x-\mu)\right]^{-(n+d)/2}$$

The result of the previous section relating normal and Student's t distributions can be extended for the multivariate case. The difference here is that two possibilities for the scale mixture form are available. In the first one, the scale of the normal distribution is altered by a scalar random quantity and in the second one it is altered by a random matrix.

A pair (x_1, x_2), where x_1 is a d-dimensional random vector and x_2 is a random variable, has multivariate normal-Gamma distribution with parameters μ, Σ, n and S, denoted $NG(\mu, \Sigma, n, S)$ or $NG_d(\mu, \Sigma, n, S)$, if $x_1|x_2 \sim N_d(\mu, \Sigma/x_2)$ and $x_2 \sim G(n/2, nS/2)$. This distribution is a multivariate extension of the normal-Gamma and also follows that $x_1 \sim t_n(\mu, S\Sigma)$.

A pair (x_1, X_2), where x_1 is a d-dimensional random vector and X_2 is a random matrix, has multivariate normal-Wishart distribution with parameters μ, σ^2, n and S, denoted $NW(\mu, \Sigma, n, S)$ or $NW_d(\mu, \Sigma, n, S)$, if $x_1|X_2 \sim N_d(\mu, \sigma^2 X_2^{-1})$ and $X_2 \sim W_d(n/2, nS/2)$. This distribution is another multivariate extension of the normal-Gamma and also follows that $x_1 \sim t_n(\mu, \sigma^2 S)$.

The above results provide two ways of generating multivariate Student's t variates via a scale mixture of normals. In the first one, a quantity x_2 is drawn from a $G(n/2, n/2)$ distribution and a vector x_1 is generated from a $N(\mu, \Sigma/x_2)$. In the second one, a matrix X_2 is drawn from a $W(n/2, n\Sigma/2)$ distribution and a vector x_1 is generated from a $N(\mu, X_2^{-1})$. Both cases will require a (Choleski) decomposition of Σ. The first method needs one generation from a Gamma distribution and d normal generations while the second method, even when the efficient technique of Odell and Feiveson is used, needs d Gamma generations and $d + d(d-1)/2 = d(d+1)/2$ normal generations. There is a clear computational advantage of the first method.

1.5 Resampling methods

Resampling methods can generically be described as generation techniques that require sampling of random variables in more than one step. Typically, they consist of two steps with the first one providing a value sampled from an approximating distribution. The second step is some sort of correction mechanism whose function is to redirect the sample so as to make it (at least approximately) representative of the distribution of interest. The correction mechanism is typically but not necessarily stochastic. In what follows, the density of interest will be denoted by π and the auxiliary density denoted by q. Unless otherwise stated, the generated values x can be scalars, vectors, matrices or even higher-dimensional arrays.

1.5.1 Rejection method

This method uses an auxiliary density for generation of random quantities from distributions not amenable to analytic treatment. More specifically, consider the generation of values from a density π for which none of the above methods provides a solution. Consider also an auxiliary density q from which draws can be made. The idea is to use q to make generations from π. The only mathematical restriction over q is that there must exist a constant $A < \infty$ such that $\pi(x) \leq Aq(x)$, for every possible value of x. For that reason, q is usually referred to as a blanketing density or an envelope and A is the envelope constant.

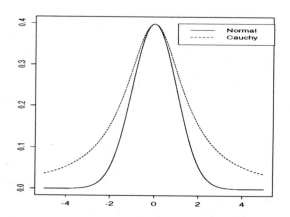

Figure 1.1. *Enveloping the $N(0,1)$ density by the $t_1(0,2.5)$ (Cauchy) density. Note that an increased scale of the envelope was needed to ensure complete blanketing.*

The method is general enough to generate values from π without knowing the complete expression for π. Generation from partially specified densities were already presented for the ratio-of-uniforms method. As will be seen throughout this book, it is extremely common in Statistics to encounter such situations where the kernel of π is known but the constant ensuring it integrates to 1 cannot be obtained analytically. Rejection methods generate from π by blanketing it with Aq irrespective of the scale used. If one is working with $\pi^*(x) = k\pi(x)$ for some constant $k = \int \pi^*(x)dx$, one would simply need to specify the envelope constant $A^* = kA$ instead of A to ensure a complete envelope. Figure 1.1 shows the blanketing of the standard normal density by a Cauchy density.

The method consists of independently drawing x from q and $u \sim U[0, 1]$ and accepting x as a value generated from π if $Auq(x) \leq \pi(x)$. Otherwise, x is not accepted as a value from π and the process must be reinitialized until a value x is accepted. Hence the name of the method, which is also known as the acceptance/rejection method. The quantity to be generated can have any form: scalar, vector or matrix. In each case, the rejection step is based on a comparison of densities with the aid of a scalar uniform variate u.

Example 1.2 A simple use of rejection is sampling from a complicated density π which is non-zero over the interval $[a_1, a_2]$ with $a_1 < a_2$ and 0 otherwise. Also, assume that π is bounded by a constant $b > 0$. An envelope is provided by the $U[a_1, a_2]$ distribution with density $q(x) = 1/(a_2 - a_1)$, for $x \in [a_1, a_2]$, and 0 otherwise. Total blanketing is ensured if $Aq(x) \geq b$, for all possible values of x and for simplicity assume $Aq(x) = b$. Hence, a draw x from π is obtained by:

1. drawing $x \sim U[a_1, a_2]$ and $u \sim U[0, 1]$;

2. accepting x if $u \leq \pi(x)/b$ and rejecting it otherwise.

This method was used, for example, to sample from the fourth component in the four-part mixture decomposition of the normal distribution described in Example 1.1 (see also Exercise 1.8e). Note that this idea can be applied when using rejection techniques for d-variate generation. In this case, π is non-zero over a region $\prod_{i=1}^{d}[a_{i1}, a_{i2}]$ and q is the d-dimensional uniform distribution over the above region.

The proof that the rejection procedure effectively generates values from π simply requires one to show that the conditional density of $[x|Auq(x) \leq \pi(x)]$ is π. The joint density of x, u is $f(x, u) = f(x)f(u) = q(x)$ by the independence between x and u and uniformity of u. Applying Bayes' theorem,

$$
\begin{aligned}
f(x|Auq(x) \leq \pi(x)) &= \frac{Pr(Auq(x) < \pi(x)|x)f(x)}{\int Pr(Auq(x) \leq \pi(x)|x)f(x)dx} \\
&= \frac{[\pi(x)/Aq(x)]q(x)}{\int [\pi(x)/Aq(x)]q(x)dx} \\
&= \frac{\pi(x)}{\int \pi(x)dx}
\end{aligned}
$$

The required density is given by the normalized version of π. So, the complete knowledge of the integrating constant of π is not required; only the kernel is needed. When π is already the complete expression of a density, the normalization is not needed as $\int \pi(x)dx = 1$.

So far nothing has been said about the choice of q. Obviously, it must be a density that is easy to draw values from. Otherwise, one is back to the original problem of generating from a difficult density. The rejection step states that x is accepted if $u < \pi(x)/Aq(x)$. When q is close to π, the envelope constant must be only slighty larger than 1 to ensure a complete envelope. In the limit, as π and q are the same, $A = 1$. On the other hand, if $q(x)$ is very different from $\pi(x)$ for some values of x, A will have to be substantially larger than 1 to ensure an envelope. In any case, the closer q and π are, the smaller the value of A and $\pi(x)/Aq(x)$ will be close to but smaller than 1. This will lead to a more likely acceptance of the proposed value. So, the efficiency of the method is directly related to similarity between π and q.

The (overall) acceptance probability is

$$
\begin{aligned}
Pr(Auq(x) < \pi(x)) &= \int Pr(Auq(x) < \pi(x)|x)q(x)dx \\
&= \int \frac{\pi(x)}{Aq(x)}q(x)dx \\
&= \frac{1}{A}\int \pi(x)dx
\end{aligned}
$$

From this expression, A must be chosen as close as possible to $\int \pi(x)dx$, which equals 1, if π is already normalized. Also, smaller values of A lead to larger acceptance probabilities. So A must be as small as possible and the only limitation is that it provides an envelope.

Great care must be exercised for the choice of A, especially when sampling from higher-dimensional densities. Often, two densities are similar in the region concentrating the bulk of the probability but their tail behaviours are very different. This may lead to very large values of A being required for total blanketing. When obtaining A numerically, such dangers that are hidden in the tails may go unnoticed and result in a constant that does not really provide an envelope. This unfortunately leads to a dilemma where a total envelope can only be achieved at the expense of very low acceptance probabilities. A more practical approach is to use a smaller constant that envelopes π only with high probability. This approach is only suitable for sampling in a subset of the space and should only be used after some analytic experimentation has ensured a very large probability for this subset.

A limiting case of rejection sampling is given for truncated distributions. Let q be any density and π its truncation to the region C. Formally, x has density $\pi(x) = kq(x)$, $x \in C$ where $k^{-1} = \int_C q(x)dx$ and 0 otherwise. Taking $A = k$ and q as the envelope density leads to accepting a generated value from q if $u < \pi(x)/kq(x)$ which is 1 if $x \in C$ and 0 if $x \notin C$. Hence, to draw a value x from q restricted to C, one simply has to draw from q and accept it if and only if it is in C, rejecting it otherwise. This simple result is heavily used in the ratio-of-uniforms method. Recall from section 1.3 that the method is based on sampling uniformly from a region $C_f \in [0, a] \times [b_1, b_2]$. To do that, one simply has to draw from a bivariate uniform distribution on $[0, a] \times [b_1, b_2]$ and retain values belonging to C_f. In some cases, even tighter bounds on C_f can be found but the truncation idea remains the same.

Example 1.3 Cheng and Feast (1979) consider the generation of $y \sim G(\alpha, 1)$ quantity, $\alpha > 1$, by the ratio-of-uniform method. They showed that the region C_f is given by $\{(x_1, x_2) : 0 < x_1^2 < (x_2/x_1)^{\alpha-1} e^{-x_2/x_1}\}$ and that $C_f \subset [0, a] \times [0, b]$ where $a = [(\alpha - 1)/e]^{(\alpha-1)/2}$ and $b = [(\alpha + 1)/e]^{(\alpha+1)/2}$. After drawing $x_1 \sim U[0, a]$ and $x_2 \sim U[0, b]$ independently, the method reduces to accepting $y = x_2/x_1$ if and only if $2 \log x_1 - (\alpha - 1) \log y + y \leq 1$ and rejecting it otherwise. Tighter bounds to C_f can be found when $\alpha > 2.5$, which leads to the more efficient algorithm:

1. draw $u_1 \sim U[0, 1]$ and $u_2 \sim U[0, 1]$ independently and set $c = [\alpha - (1/6\alpha)]$;

2. if $\alpha > 2.5$, reset $u_1 = u_2 + (1 - 1.86u_1)/\sqrt{\alpha}$ until $0 < u_1 < 1$;

3. accept $y = cu_2/u_1$ if $2 \log u_1 - (\alpha - 1) \log y/(\alpha - 1) + y \leq 1$ and reject it otherwise.

The case $\alpha < 1$ was tackled by Ahrens and Dieter (1974). Their algorithm is:

1. draw $u_1 \sim U[0, 1]$ and $u_2 \sim U[0, 1]$ independently and set $c = 1 + \alpha/e$;

2. if $u_1 \leq 1/c$, set $x = (cu_1)^{1/\alpha}$ and $y = e^{-x}$, otherwise set $x = -\log[c(1 - u_1)/\alpha]$ and $y = x^{\alpha-1}$;

3. accept x if $u_2 < y$ and reject it otherwise.

When $\alpha < 1$, another sampling scheme using the algorithm for $\alpha > 1$ is to take $u \sim U[0, 1]$ and $y \sim G(\alpha + 1, 1)$ independently and take $x = yu^{1/\alpha}$ (Exercise 1.16). Whatever the value of α, once a draw x from a $G(\alpha, 1)$ is obtained, $x/\beta \sim G(\alpha, \beta)$ (Exercise 1.5). Many other algorithms for gamma generation have been proposed (Ripley, 1987).

If computation of π is costly then some other auxiliary function $s(x) \leq \pi(x)$ can be used as a preliminary test (or pretest). Note that π gets squeezed between s and q. In this case, before verifying acceptance, one verifies whether $u < s(x)/Aq(x)$. As $s(x)/Aq(x) \leq \pi(x)/Aq(x)$, the method will indicate acceptance of x without needing evaluation of $\pi(x)$. Only in the negative case, the method proceeds as before. Again, if the pretest is efficient (s is close to π) and the rejection method is efficient (q is close to π) then π will need to be evaluated only a few times. This procedure is called pretesting or squeezing (Marsaglia, 1977).

1.5.2 Weighted resampling method

Determination of the constant A is vital for the use of rejection methods. As already shown, this is by no means an easy task in general. Weighted resampling techniques are based on borrowing the good idea of sampling from an approximating density q without having to find the constant A. The disadvantage of the method is that it only provides values approximately distributed according to π.

Assuming the presence of a sample $x_1, ..., x_n$ from q, weights

$$w_i = \frac{\pi(x_i)/q(x_i)}{\sum_{j=1}^{n} \pi(x_j)/q(x_j)}, \quad i = 1, ..., n$$

are constructed. A second sample of size m is drawn from the discrete distribution on $\{x_1, ..., x_n\}$ with probabilities $w_1, ..., w_n$. The resulting sample has approximate distribution π. It is worth noting that here again the complete expression of π is not needed. If only $\pi^*(x) = k\pi(x)$ is known without knowledge of the normalizing constant k, the weights w_i do not alter. This method was initially proposed by Rubin (1987) in a comment and later discussed in more detail by Rubin (1988) and can be applied to quantities x of any dimension. The example considered by Rubin (1987) suggested that q be taken as a product of univariate normal densities for the components of x, after transforming them to take values on the real line. So, a logarithmic operation would be imposed over positive quantities. In the next chapters, better approximations will be presented.

To verify that the method generates approximately from π, consider the distribution function of a univariate x (for ease of exposition) generated by the algorithm

$$F_x(a) = Pr(x \leq a) = \sum_{\{i:x_i \leq a\}} w_i = \frac{\sum_{i=1}^{n} \pi(x_i)/q(x_i)I(x_i \leq a)}{\sum_{i=1}^{n} \pi(x_i)/q(x_i)}.$$

Under appropriate regularity conditions usually satisfied in practice, when $n \to \infty$, a large quantity of x_i are available with frequencies according to q, and sums become integrals weighted by q. Therefore,

$$F_x(a) \to \frac{\int [\pi(x)/q(x)]I(x \leq a)\,q(x)\,dx}{\int [\pi(x)/q(x)]\,q(x)\,dx} = \frac{\int \pi(x)I(x \leq a)\,dx}{\int \pi(x)\,dx}$$

that is the distribution function associated with the density π. The derivation above can be extended with more cumbersome notation to the case of a vector or matrix x. It is clear from above that if the initial sample is sufficiently large the resulting sample will have aproximate distribution π.

As for the rejection method, q must reflect as much as possible the characteristics of π. Otherwise, the sample will be an inappropriate basis for resampling, compromising the method. This point is better understood by referring to Figure 1.2 with a bad choice for the starting point q in the univariate case. A sample from q, also marked in the figure, is almost entirely concentrated in regions where π is negligible and, worse still, is almost absent from the region where π is relevant. As a result, the resample will be a very bad representation of the distribution of interest unless the initial sample size n is extremely large. For high dimensional problems, this is virtually impossible to attain. In summary, q should be as similar to π as possible.

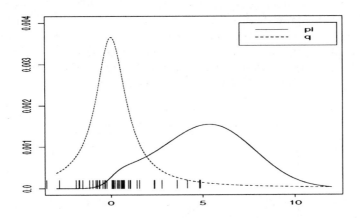

Figure 1.2. *Example of an inadequate choice of initial density q. The small bars along the x axis represent points drawn from q.*

The weighted resample can have any size, unlike the rejection method where the resulting sample is necessarily smaller than the original one. The resample can be drawn with or without replacement without affecting the results provided the initial sample n is of order of magnitude larger than the resulting sample m. The most important requirement is to have enough initial points to cover adequately the regions of interest of π. Rubin (1987, 1988) points out that the ratio n/m should be decreased towards 1 as q and π get similar and suggests taking $n/m = 20$. He uses the name sampling/importance resampling (SIR, in short) in analogy with Monte Carlo integration (section 3.5) where sampling-based approximations of integrals may be improved by an appropriate choice of the sampling density. Weighted resampling is a more concise description although the preferred nomenclature is a matter of personal taste.

1.5.3 Adaptive rejection method

The analytic form of the density in complex situations, to be presented in the sequel, virtually prevents the establishment of envelopes by simple study of their mathematical expressions. Nevertheless, these envelopes may be constructed with the aid of graphical techniques. This is certainly viable for the important and encompassing class of log-concave densities. A density $\pi(x)$ is log-concave if the vector of the derivative of its logarithm exists and has components that are non-increasing in x. Another definition is to have a non-positive determinant of the matrix of second derivatives of its logarithm for all values of x. Log-concavity can also be defined without using the derivatives. For the univariate case, π is log-concave if $\log \pi(x_1) - \log \pi(x_2) \leq \log \pi(x_2) - \log \pi(x_3)$ for $x_1 < x_2 < x_3$ or $d \log \pi / dx$ is a non-increasing function of x or $d^2 \log \pi / dx^2$ is a non-positive function of x. Many densities are log-concave including the normal, the Gamma ($\alpha > 1$), the Wishart ($\alpha > (p+1)/2$) and the densities arising in the context of generalized linear models with canonical link (Dellaportas and Smith, 1993). A list with log-concavity checks for the main distributions is given by Gilks and Wild (1992, Table 2).

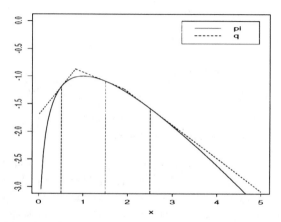

Figure 1.3. *Adaptive envelope of a log-concave density. The G(2,1) density is graphed above in the log scale with a starting grid $\{0.5, 1.5, 2.5\}$ of size $n = 3$. The envelope q is provided by the tangent lines at the grid points. Lower bounds can be provided by joining the density at these points.*

Log-concavity can be explored to construct easy envelopes graphically. For the univariate case, Gilks and Wild (1992) developed a method based on an approximation of the logarithm of π by line segments that blanket the density from above (and also from below if squeezing techniques are also applied). Simple forms of envelopes are obtained by intersecting the tangent or secant lines at pre-chosen points $x_1, ..., x_n$. Lower bounds are provided by the line segments connecting successive points and can be used for pretesting. These are illustrated in Figure 1.3 where a reasonable approximation was obtained with the number of points as small as 3 and it improves as the number of points used increases. The use of tangents provides a better approximation than the use of secants but the only obvious choice for lower bounds is provided by the chords. The disadvantage of tangent lines is the need to evaluate the derivatives of $\log \pi$ (or of π) at points $x_1, x_2, ..., x_n$. The envelope based on secant lines only requires the evaluation of π at these points (Gilks, 1992). Therefore, the choice between tangent or secant lines has to do with the computational price of evaluation of the derivatives of π.

In any case, the function enveloping $\log \pi$ is a succession of lines with different inclination and hence the function enveloping π is a succession of exponentials. The density q can be taken as the density of the piecewise exponential distributions with parameters given by the derivatives of $\log \pi$ at points $x_1, ..., x_n$ (in the tangent case) and partition determined by the intersection points of adjacent lines. It is not difficult to draw values from this distribution (Exercise 1.18) and an envelope is obtained graphically. Wild and Gilks (1993) detail the algorithm and a Fortran code is available from these authors.

The adaptation referred to in the name of this method has to do with the fact that the envelopes can be increasingly improved as more points are added to the grid $\{x_1, ..., x_n\}$. With such an initial grid, a piecewise exponential distribution is obtained for q and a value x^* drawn from it. If the value is accepted, the process is terminated and x^* is a value sampled from π. If the value is rejected, it can be incorporated to the *current* grid. If x^* belongs to $(x_j, x_{j+1}]$ for $1 \leq j < n - 1$, a new tangent line is incorporated to the envelope at that point. As the size of the grid increases, the approximation of q to π becomes better and rejection becomes less likely.

In general 2 to 4 points are enough for the starting grid and on average the same number of points are drawn before a value is accepted. Log-concavity is essential to ensure that a succession of tangent or secant lines provides an envelope and the existence of an envelope is a condition for the rejection method. In Chapter 6, a variation of the rejection method that does not require the approximating density to blanket the target density completely is presented. Gilks, Best and Tan (1995) used this idea to construct an adaptive sampling method for cases where the density of interest is not log-concave and hence the succession of lines does not envelope it entirely.

The technique of constructing an envelope with tangent lines used by the adaptive rejection method in the univariate case may be extended to tangent planes (and hyperplanes) in the bivariate (multivariate) case. The resulting density obtained by intersecting the planes will still be an envelope. The main problem is the difficulty in sampling from this envelope, making this technique very unattractive for use beyond the univariate case.

1.6 Exercises

1. Show that reordering the possible values x_i of a discrete random quantity

x according to a decreasing order of their respective probabilities speeds up generation of x by reducing the average number of comparisons.

2. Describe a scheme to draw values from the following discrete distributions:

   (a) multinomial;

   (b) hypergeometric.

3. Prove that $u \sim U[0,1]$ if and only if $1 - u \sim U[0,1]$.

4. Obtain the distribution function of the Weibull distribution and use the probability integral transform to confirm that $[-(1/\lambda) \log u]^{1/\alpha}$ where $u \sim U[0,1]$.

5. x admits a location model if its density has the form $f(x|a) = f_0(x-a)$, admits a scale model if its density has the form $f(x|b) = (1/b)f_0(x/b)$ and admits a location-scale model if its density has the form $f(x|a,b) = (1/b)f_0[(x-a)/b]$ where $f_0(x)$ is a density depending only on x.

   (a) Show that generation of x from a quantity x_0 generated from f_0 is given by $x = x_0 + a$ in the location model, $x = bx_0$ in the scale model and $x = bx_0 + a$ in the location-scale model.

   (b) Apply the results in (a) to draw samples from the $U[a-1/2, a+1/2]$, $U[0,b]$ and $U[a-b, a+b]$ distributions based on samples from the $U[0,1]$ distribution.

6. Let u_1 and u_2 be independent $U[0,1]$ variates, $x_i = 2u_1 - 1$, $i = 1,2$ and $w = u_1^2 + u_2^2$. If $w > 1$, a new pair (u_1, u_2) should be considered. If $w \leq 1$, define $y_i = \sqrt{-2(\log w)/w}\, x_i$, $i = 1,2$. Show that y_1 and y_2 are independent $N(0,1)$ quantities.

7. Use the mixture technique to draw values from the density $f(x) = 1.5(1 - x^2)$, $0 < x < 1$, using as approximation $f_1(x) = 2(1 - x)$, $0 < x < 1$, and 0 otherwise. Obtain the value of a_1 and the expression of the residual density $g_1(x)$. Propose alternative sampling schemes and discuss advantages and disadvantages of the mixture technique for this particular case.

8. (Marsaglia and Bray, 1964) A standard normal quantity x may be generated using mixtures. Assume $x_1, ..., x_7$ are independent $U[0,1]$ and consider components $x_1 = 2(x_1 + x_2 + x_3 - 1.5)$, $x_2 = 1.5(x_4 + x_5 - 1)$, x_3 providing the value of x when $|x| > 3$ using the method of Box and Miller and rejection and x_4 being the component that completes the mixture. Each component x_i has density f_i and probability a_i, $i = 1, 2, 3, 4$.

(a) Show that x_1 and x_2 have densities

$$f_1(x) = \begin{cases} (3 - x^2)/8 & , \quad |x| \leq 1 \\ (3 - |x|)^2/16 & , \quad 1 < |x| < 3 \quad \text{and} \\ 0 & , \quad \text{otherwise} \end{cases}$$

$$f_2(x) = \begin{cases} (6 - 4|x|)/9 & , \quad |x| \leq 1.5 \\ 0 & , \quad \text{otherwise} \end{cases}$$

(b) Show (analytically) that the largest weight component x_1 can have is $a_1 = 0.8638$.

(c) Show (numerically or graphically) that the largest weight component x_2 can have is $a_2 = 0.1107$.

(d) Show that component x_3 has weight $a_3 = 0.0027$ and describe how to implement a draw from x_3.

(e) Show that component x_4 has weight $a_4 = 0.0228$ and can be generated by taking the first value of x_6 such that $f_4(6u_6 - 3) \geq 0.3181u_7$ where

$$f_4(x) = \frac{f(x) - p_1 f_1(x) - p_2 f_2(x) - p_3 f_3(x)}{1 - p_1 - p_2 - p_3}$$

(f) Certify yourself of the correctness of the algorithm by writing a computer program, drawing a sample of size 1000 and plotting the resulting histogram along with the standard normal density.

(g) Discuss advantages and disadvantages of the above algorithm with respect to the method of Box and Miller.

9. (Wakefield, Gelfand and Smith, 1991) Consider the generation of a pair (x_1, x_2) uniformly distributed in the region $C_{f,g} = \{(x_1, x_2) : 0 \geq x_1 \geq g^{-1}[cf^*(x_2/g'(x_1))]\}$ with a strictly increasing differentiable function g and a constant $c > 0$.

(a) Show that $y = x_2/g'(x_1)$ has a distribution with density $f(y) = f^*(y)/\int f^*(u)du$.

(b) Consider the power family $g(x) = x^{r+1}/(r + 1)$ for $r \geq 0$ and $c = 1/(r + 1)$. Show that if $f^*(x)$ and $x^{r+1}[f^*(x)]^r$ are bounded then the constants $a = \sup_x [f^*(x)]^{1/(r+1)}$, $b_1 = \inf_{x \leq 0} x[f^*(x)]^{r/(r+1)}$ and $b_2 = \sup_{x \geq 0} x[f^*(x)]^{r/(r+1)}$ are well defined and $C_{f,g} \in [0, a] \times [b_1, b_2]$.

(c) Show for the conditions of (b) that the acceptance probability of a point uniformly generated in $[0, a] \times [b_1, b_2]$ is given by

$$\frac{\int f^*(x)dx}{(r + 1)a(b_2 - b_1)}$$

(d) Obtain for the normal case $(f^*(x) = e^{-x^2/2})$ that $a = 1$, $b_2 = -b_1 = \sqrt{(r + 1)/re}$ and that the acceptance probability in (c) is maximized for $r = 0.5$ and its value is 0.731.

(e) Consider the generation of a vector $(x_1, ..., x_d, x_{d+1})'$ uniformly distributed in the region $C_{f,g} = \{(x_1, ..., x_d, x_{d+1})' : 0 \geq x_{d+1} \geq [f^*(y_1, ..., y_d)]^{1/(rd+1)}\}$, with $y_i = x_i/x_{d+1}$, $i = 1, ..., d$. Show that $y = (y_1, ..., y_d)'$ has a multivariate distribution with joint density $f(y) = f^*(y)/ \int f^*(u)du$.

(f) Show that if $f^*(x)$ and $x_i^{rd+1}[f^*(x)]^r$ are bounded with $x = (x_1, ..., x_d)'$ then the constants $a = \sup_x [f^*(x)]^{1/(rd+1)}$, $b_{1i} = \inf_{x_i \leq 0} x_i[f^*(x)]^{r/(rd+1)}$ and $b_{2i} = \sup_{x_i \geq 0} x_i[f^*(x)]^{r/(rd+1)}$ are well defined and $C_{f,g} \in \prod_{i=1}^{d} [b_{1i}, b_{2i}] \times [0, a]$.

10. Describe 3 different methods for generation of random quantities with Cauchy distribution of density

$$f(x) = \frac{1}{\pi} \frac{1}{1 + x^2} \quad , \quad x \in R$$

with the first method using probability integral transform, the second one using the ratio-of-uniforms from region $\{(x_1, x_2) : x_1 > 0$ and $x_1^2 + x_2^2 < 1\}$ and a third one using scale mixture of normals.

11. Show that if a pair (x_1, x_2) has distribution:

(a) $NG(\mu, \Sigma, n, S)$ then $x_1 \sim t_n(\mu, S\Sigma)$.

(b) $NW(\mu, \sigma^2, n, S)$ then $x_1 \sim t_n(\mu, \sigma^2 S)$.

12. x has double exponential distribution with parameters μ and σ, denoted by $DE(\mu, \sigma)$, if its density is

$$f(x|\mu, \sigma) = \frac{1}{2} \exp \left\{ -\frac{|x - \mu|}{\sigma} \right\}, \text{ for } x \in R$$

(a) Using the notation of Exercise 1.5, show that the DE distribution admits a location-scale model with location parameter μ and scale parameter σ.

(b) Show that x can be obtained as a discrete mixture of y and $-y$ with equal weights where $y \sim Exp(1)$ and describe a generation scheme for x based on (a) and the above result.

(c) Show that if $x|y \sim N(0, y)$ and $y \sim Exp(1/2)$ then $x \sim DE(0, 1)$ and describe a generation scheme for x based on (a) and the above result.

(d) Compare the generation schemes described in (b) and (c).

13. (Odell and Feiveson, 1966) Let $Y = LL'$ where L is a $d \times d$ lower triangular matrix. Show that if the main diagonal elements l_{ii} of L are square roots of $G((n - i + 1)/2, 1/2)$ (or χ^2_{n-i+1}) quantities, the off-diagonal elements l_{ij} of L are $N(0, 1)$ quantities and all these variates are independently drawn then $Y \sim W_d(n/2, I_d/2)$.

14. (Cheng and Feast, 1979) Consider the generation of an $x \sim G(\alpha, 1)$ quantity, $\alpha > 1$, by the ratio-of-uniform method. Show that:

(a) the region C_f is given by $\{(x_1, x_2) : 0 < x_1^2 < (x_2/x_1)^{\alpha-1} e^{-x_2/x_1}\}$;

(b) the region $C_f \subset [0, a] \times [0, b]$ where $a = [(\alpha - 1)/e]^{(\alpha-1)/2}$ and $b = [(\alpha + 1)/e]^{(\alpha+1)/2}$;

(c) after drawing $x_1 \sim U[0, a]$ and $x_2 \sim U[0, b]$ independently, the method reduces to accepting $y = x_2/x_1$ if and only if $2 \log x_1 - (\alpha - 1) \log y + y \le 1$ and rejecting it otherwise;

(d) tighter bounds to C_f can be found when $\alpha > 2.5$;

(e) using (d) a more efficient algorithm is given by:

    i. draw $u_1 \sim U[0, 1]$ and $u_2 \sim U[0, 1]$ independently;

    ii. if $\alpha > 2.5$, reset $u_1 = u_2 + (1 - 1.86u_1)/\sqrt{\alpha}$ until $0 < u_1 < 1$;

    iii. accept $y = cu_2/u_1$ if $[2 \log u_1 - (\alpha - 1) \log y/(\alpha - 1) + y] \le 1$ and reject it otherwise with $c = [\alpha - (1/6\alpha)]$.

15. Describe in detail the generation of a value from a $G(\alpha, \beta)$ using the rejection method with $q = G(n, c)$, n integer. Answer in particular:

(a) What values of n and c should be chosen?

(b) Once values for n and c are chosen, what is the value of the constant A that optimizes the method?

16. Show that to generate $x \sim G(\alpha, 1)$ with $\alpha < 1$ using the algorithm for $\alpha > 1$ one can take $u \sim U[0, 1]$ and $y \sim G(\alpha + 1, 1)$ independently and make $x = yu^{1/\alpha}$. Write a computer program to draw samples from a $G(\alpha, 1)$ distribution using this and Ahrens and Dieter algorithms for a variety of values of α. Compare the two algorithms.

17. Suppose one wishes to sample from a $N(3, 1)$ distribution using the SIR algorithm. There is previous information allowing the specification of scale parameter to 1 but little knowledge about the location. Therefore the location of importance density is taken as 0 and the heavy-tailed Cauchy distribution is chosen to accommodate this uncertainty. In order to form a representative sample of size $m = 1000$ from the distribution of interest, one would expect to have around 477 (preferably different) values between 3 and 5. How large should an initial sample size n be to allow one to expect this many values on the interval $[3, 5]$?

18. x has piecewise exponential distribution with parameters $\lambda_1, ..., \lambda_n$ and partition $t_1, ..., t_n$ if its density is

$$f(x) = k_i \lambda_i e^{-\lambda_i(x - t_{i-1})}, \ x \in (t_{i-1}, t_i], \ i = 1, ..., n$$

where $t_0 = 0$, $k_1 = 1$ and

$$k_i = \exp\left[-\sum_{j=1}^{i-1}\lambda_j(t_j - t_{j-1})\right], \quad i = 2,...n$$

(a) Show that $f(x|x > t_{i-1}) = \lambda_i\,e^{-\lambda_i(x-t_{i-1})}$, $x \in (t_{i-1}, t_i]$ and, therefore, $\min(x - t_{i-1}, t_i - t_{i-1})|x > t_{i-1} \sim Exp(\lambda_i)$, $i = 1,...,n$.

(b) Use the result in (a) to sample values from this distribution.

19. Show with and without evaluating derivatives that the univariate and multivariate normal, the Gamma ($\alpha > 1$) and the Wishart ($\alpha > (p + 1)/2$) densities are log-concave.

## CHAPTER 2

Bayesian inference

2.1 Introduction

The concepts involved in the Bayesian approach to inference are described in this chapter. The treatment and presentation are not exhaustive. The main interest is to show how the ingredients required for the construction of a model are formulated in a given problem and what techniques are used for understanding and extracting relevant information from the process. For a more thorough discussion of Bayesian inference, the reader is referred to the books by Berger (1985), Gamerman and Migon (1993), O'Hagan (1994), Robert (1994) and Smith and Bernardo (1994). A more philosophical approach is given by De Finetti (1974, 1975).

The techniques will be derived in more detail for classes of models to be considered in subsequent chapters. In particular, emphasis will be placed on linear regression models and their many generalizations, which include generalized linear models, hierarchical models, dynamic linear models and dynamic generalized linear models.

In the next section the standard Bayesian point of view is presented. It consists of the combination of historic and data information through Bayes' theorem and the resulting consequences. In section 2.3, a special family of distributions that plays an important role in the specification of models will be introduced: the conjugate family of distributions. The use of these distributions is illustrated in the context of normal regression models. In the next sections, the hierarchical models and the dynamic models are described. Inference for the case of normal observations and the generalized case of non-normal observations are presented in both cases.

2.2 Bayes' theorem

As in the classical or frequentist approach to inference, the Bayesian approach is developed in the presence of observations x whose value is initially uncertain and described through a probability distribution with density or probability function* $f(x|\theta)$. The quantity θ serves as an index of the family

---

* As will be seen in the sequel, the mathematical treatment can be unified without differentiation between discrete and continuous quantities. The terminology of continuous variables is chosen for simplicity and f will be referred to as a density. Whenever required, a differentiation is made in the text.

of possible distributions for the observations. It represent characteristics of interest one would wish to know in order to obtain a complete description of the process.

The canonical situation is one where a random sample $x = (x_1, ..., x_n)'$ is extracted from a population that is distributed according to the density $f(x|\theta)$. Typically in this case, the observations x_i are identically distributed and independent (conditionally on the knowledge of θ).

Example 2.1 Consider a series of measurements about a physical quantity θ with measurement errors e_i described by the $N(0, \sigma^2)$ distribution, where the precision of the measurements (controlled by σ^2) is known. In this case, $x_i = \theta + e_i$, $i = 1, ..., n$ and

$$f(x|\theta) = \prod_{i=1}^{n} f_N(x_i; \theta, \sigma^2) = \prod_{i=1}^{n} \frac{1}{\sqrt{2\pi\sigma^2}} \exp\left\{ -\frac{1}{2} \frac{(x_i - \theta)^2}{\sigma^2} \right\}$$

It is important to note that the quantity θ is more than a simple index. It is the very reason for drawing measurements as the main interest of the study is to determine its value.

2.2.1 Prior, posterior and predictive distributions

The situation of Example 2.1 repeats in more general cases, that is, the index θ is a quantity of interest with a very precise meaning to the problem under study. Furthermore, it is likely that the researcher has some knowledge about its value. It is then possible and even scientifically recommended that this body of knowledge should be formally incorporated in the analysis. At this point, the Bayesian and frequentist approaches diverge. The second one does not admit this information because it has not been observed and is therefore not subject to empiric verification. The Bayesian approach incorporates this information to the analysis through a density $p(\theta)$ even when this information is not precise.

There used to be much controversy about the appropriateness of incorporating this historic information in the analysis; see Efron (1986), Lindley (1978) and Smith (1984) for a sample of the discussion about the theme. Nowadays, the dominant impression is that this discussion is no longer relevant, with the difference between the two approaches well understood and elements of each one used wherever it is more suitable to do so. The ultimate test is applicability of the theory and the Bayesian approach used to fare badly on this account due to the inclusion of more elements into the analysis. The techniques described in this text allow the use of Bayesian inference in quite complex problems, virtually removing its main restriction. Therefore, from now on this text will assume firm adherence to the Bayesian paradigm and this chapter is devoted to presenting its basic elements.

As described above, Bayesian inference contains two ingredients: the observational (or sampling) distribution $f(x|\theta)$ and the distribution $p(\theta)$. This latter distribution can also be specified with the help of constants just like the distribution of x. Sometimes it is useful to distinguish them from the parameter of interest θ. These constants are then called hyperparameters, as they are the parameters of the distribution of the parameters. Initially, the hyperparameters are assumed to be known.

Looking at the first ingredient of Bayesian inference as a function of θ gives the likelihood function of θ, $l(\theta) = f(x|\theta)$. It provides the chances of each value of θ having led to that observed value of x. The second ingredient is called prior density as it contains the probability distribution of θ *before* the observation of the value of x. Once the problem is cast in this form it is only natural that inference should be based on the probability distribution of θ *after* observing the value of x, that becomes part of the set of available information. This distribution is called posterior distribution in direct opposition to prior distribution. It can be obtained by Bayes' theorem

$$p(\theta|x) = \frac{f(x|\theta)p(\theta)}{f(x)}$$

where

$$f(x) = \int f(x|\theta)p(\theta)d\theta \tag{2.1}$$

As the left hand side is a density for θ, the observation x is simply a constant as well as $f(x)$. Furthermore, the posterior will be denoted from now on by $\pi(\theta)$, with the value of the observation upon which it was conditioned implicitly understood. Bayes' theorem can then be written in a more compact form

$$\pi(\theta) \propto l(\theta)\,p(\theta) \tag{2.2}$$

The concepts of prior and posterior are always relative to the observation considered at a given moment. It is possible that after observing x and obtaining the posterior, a new observation y also related to θ through an eventually different likelihood function becomes available. In this case, the posterior (relative to x) is the prior (relative to y) and a new posterior can be obtained by a new application of Bayes' theorem. It can be easily shown that the posterior resulting from observations of x and y is the same irrespective of the order in which x and y were processed (Exercise 2.1).

Example 2.1 (continued) The model can be completed with a prior distribution $p(\theta) = N(\mu, \tau^2)$, where the hyperparameters μ and τ^2 are known constants. The likelihood function is

$$l(\theta) = \prod_{i=1}^{n} \frac{1}{\sqrt{2\pi\sigma^2}} \exp\left\{-\frac{1}{2}\frac{(x_i - \theta)^2}{\sigma^2}\right\} \propto \exp\left\{-\frac{n}{2\sigma^2}(\bar{x} - \theta)^2\right\}$$

where $\bar{x}$ is the arithmetic average of the x_i. Therefore, the posterior density is

$$\pi(\theta) \propto \exp\left\{-\frac{1}{2}\frac{(\bar{x} - \theta)^2}{\sigma^2/n}\right\} \exp\left\{-\frac{1}{2}\frac{(\theta - \mu)^2}{\tau^2}\right\}$$

$$\propto \exp\left\{-\frac{1}{2}\left[\frac{(\theta - \mu_1)^2}{\tau_1^2} + \frac{(\bar{x} - \mu)^2}{n^{-1}\sigma^2 + \tau^2}\right]\right\}$$

where $\tau_1^{-2} = n\sigma^{-2} + \tau^{-2}$ and $\mu_1 = \tau_1^2(n\sigma^{-2}\bar{x} + \tau^{-2}\mu)$. The last passage is obtained using that if x, a_1, a_2, b_1 and b_2 are scalars then

$$\frac{(z - a_1)^2}{b_1} + \frac{(x - a_2)^2}{b_2} = \frac{(z - c)^2}{d} + \frac{(a_1 - a_2)^2}{b_1 + b_2}$$

where $d^{-1} = b_1^{-1} + b_2^{-1}$ and $c = d(b_1^{-1}a_1 + b_2^{-1}a_2)$ with $z = \theta$, $a_1 = \bar{x}$, $a_2 = \mu$, $b_1 = \sigma^2/n$ and $b_2 = \tau^2$. Incorporating the multiplicative term that does not depend on θ to the proportionality constant gives

$$\pi(\theta) \propto \exp\left\{-\frac{1}{2}\frac{(\theta - \mu_1)^2}{\tau_1^2}\right\}$$

In other words, the posterior distribution of θ is $\pi(\theta) = N(\mu_1, \tau_1^2)$.

Note that by increasing the value of τ^2, the information contained in the prior is reduced and so is its influence on the analysis. In the limit when $\tau^2 \to \infty$ the non-informative prior $p(\theta) \propto k$ is obtained and $\pi(\theta) = N(\bar{x}, \sigma^2/n)$.

There is plenty of controversy among Bayesians about the specification of non-informative prior distributions. Part of the disagreement is due to an inherent anomaly of these distributions. Often, this specification leads to improper distributions. These are distributions that do not integrate to 1 as prescribed by the theory of probability. Note from the example that when $\tau^2 \to \infty$, $\int p(\theta)d\theta \neq 1$. Besides, there are many different definitions of non-informative prior distributions, especially in multivariate cases. One of the most commonly accepted definitions is Jeffreys' prior, given by $p(\theta) \propto |I(\theta)|^{1/2}$ where

$$I(\theta) = E\left[-\frac{\partial^2 \log f(x|\theta)}{\partial\theta\,\partial\theta'}|\theta\right] \tag{2.3}$$

is the expected Fisher information matrix about θ. Jeffreys (1961) formalized a theory of Bayesian inference mostly using this prior and justified it on the grounds of invariance under parametric transformations. In general, it leads to prior densities in the form $p(\theta) \propto k$ for location parameters θ, $p(\sigma) \propto \sigma^{-1}$ for scale parameters σ. When a location θ and scale σ are present, Jeffreys (1961) suggested a change from the above rule to a product rule that leads to $p(\theta, \sigma) \propto \sigma^{-1}$.

Another pre-eminent specification of vague prior information is provided by the reference approach introduced by Bernardo (1979) and later refined by Berger and Bernardo (1992) and Berger, Bernardo and Mendoza (1989). It is based on expected discrepancy measures of information and under asymptotic normality coincides with Jeffreys' prior in the univariate case. In the multidimensional case, the reference approach works on splitting the parameter vector into groups and seems to avoid some difficulties of other approaches in the multiparameter case, albeit at the cost of a more complex derivation of the prior.

The impropriety of some vague prior specifications is a nuisance but in general they lead to proper posterior distributions and inference can be made without any difficulty. There are exceptions and in some cases the posterior remains improper. This is a serious problem as in many complex models verification of propriety is far from trivial. For these models, exact inference cannot be performed and the approximations used may lead to a number of inconsistencies. The recommendation is therefore to avoid improper specifications if possible or use them with caution otherwise.

Another important element for Bayesian inference is the predictive or marginal distribution of x with density $f(x)$ given by (2.1). It provides the expected distribution for the observation x as $f(x) = E[f(x|\theta)]$ and the expectation is taken with respect to the prior distribution of θ. A similar derivation can be applied to the prediction of a future observation y after observing x. This prediction should be based on the distribution of $y|x$, that is, on the updated probabilistic description based on the available information. If y and x are conditionally independent given θ then

$$f(y|x) = \int f(y, \theta|x) d\theta = \int f(y|\theta)\, \pi(\theta)\, d\theta$$

and again the density is obtained as the expectation of the sampling distribution but this time with respect to the posterior of θ. Conditional independence between x and y is obtained, for example, if $x = (x_1, ..., x_n)'$ and $y = (x_{n+1}, ..., x_{n+m})'$ are samples from $f(x|\theta)$. The predictive distribution is then used to predict future values of this population.

Predictive distributions form the basis of the predictive approach to inference. This approach is described, detailed and applied to a variety of problems by Aitchinson and Dunsmore (1975) and Geisser (1993). The main thrust of their argument is that the ultimate test of any inferential procedure is the confrontation against reality. When making inference about parameters this will not be possible as they will not be observed. Nevertheless, it is still useful to be able to make probabilistic statements about meaningful parameters as that will have a direct effect on some course of action.

In the case of a multivariate parameter $\theta = (\theta_1, ..., \theta_d)'$, marginal and conditional distributions of the components θ_i can be obtained from the joint density $\pi(\theta_1, ..., \theta_d)$. The marginal posterior density of θ_i is

$$\pi(\theta_i) = \int \pi(\theta_1, ..., \theta_d) \, d\theta_{-i} \qquad (2.4)$$

where $\theta_{-i} = (\theta_1, ..., \theta_{i-1}, \theta_{i+1}, ..., \theta_d)$ is the vector θ with its ith component removed, $i = 1, ..., d$. For each θ_i the possible conditional distributions are

$$\pi(\theta_i | \theta_j, j \in C) = \pi(\theta_i, \theta_j, j \in C) / \pi(\theta_j, j \in C)$$

for all $C \subset \{1, ..., i-1, i+1, ..., p\}$. The most important for this text is the conditional distribution of $\theta_i | \theta_{-i}$, called full conditional distribution of θ_i and with density denoted by $\pi_i(\theta_i)$. The above presentation can be analogously extended for the case where each of the components θ_i is itself a vector.

2.2.2 Summarizing the information

Once the posterior distribution is available, one may seek to summarize its information through a few elements. In particular, location and dispersion measures may be calculated to provide an idea of possible central values and variability associated with the posterior. The main location measures are mean, mode and median and the main dispersion measures are variance, standard deviation, precision, interquartile range and curvature at the mode. The posterior mean is the expected value of θ under π, the posterior mode is the most likely value under π and the median divides the parametric space in two equal probability parts. In the multivariate case, the variance is given by a matrix and the standard deviation can be seen as the vector of square roots of the diagonal elements of this matrix. The precision is given by the inverse of the variance matrix and the curvature at the mode given by the matrix of second derivatives of $\log \pi$ at the mode and provides a local idea about the posterior dispersion. Apart from the median, all these measures can be evaluated for joint, marginal and conditional distributions. The median only makes sense for univariate distributions. A description of these measures and their relation to decision rules is given by Gamerman and Migon (1993).

In multivariate spaces, marginal densities are useful for concentrating inference on a component of the parameter space. These are also obtained by integration (Equation (2.4)). Marginal distributions are particularly useful when summarizing the information about parameters through credibility (or probability) intervals. C is a $100(1-\alpha)\%$ credibility interval for a scalar parameter θ if $\int_C \pi(\theta) d\theta = 1 - \alpha$. Again, an integration is required. For a given value of α, the interval C of shortest length is given by the interval that includes points of higher posterior density than points not included in the interval. These intervals are called highest posterior density (HPD) intervals. The idea of probability intervals can be extended to parameters of higher dimension leading in general to credibility regions constructed in exactly the same way.

It is clear that most summarization procedures involve integration of π. Exact inference will only be possible if these integrations can be performed analytically. Otherwise, approximations will have to be used. For the remainder of this chapter, a few basic situations where integration is possible are presented. These are followed by models of practical interest where exact inference is no longer possible. They will be used as motivating examples for the approximate solution to inference via Markov chain simulations introduced later in the book.

Example 2.1 (continued) As $\pi = N(\mu_1, \tau_1^2)$, it is easy to see that the posterior mean, mode and median of θ are given by μ_1. As for posterior dispersion measures, variance and curvature at mode are given by τ_1^2, standard deviation by τ_1 and dispersion by τ_1^{-2}. In this case, in addition to the easy calculation there is an agreement between the different information indicators and summarization is straightforward. Figure 2.1 shows this density with specific values for μ, τ^2, $\bar{x}$ and σ^2/n.

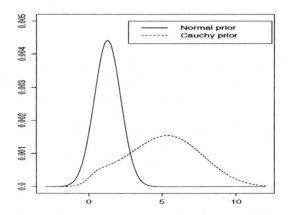

Figure 2.1. *Posterior densities obtained for Examples 2.1 and 2.2 with $\mu = 0$, $\tau^2 = 1$, $\sigma^2/n = 4.5$ and $\bar{x} = 7$. The change in the form of the prior causes major disagreement between the resulting posterior densities. While for Example 2.1 it is a well behaved $N(1.27, 0.82)$ density, for Example 2.2 it is an assymetric density with a hump to the left of the mode. Approximate location values are: mean=5.05, mode=5.36, median=5.10 and approximate dispersion values are: variance=6.10 and inverse curvature at mode=6.27.*

The results of the example are useful but unfortunately are not the rule in Bayesian inference. In general, the expression of the posterior even in simple models is complex enough not to allow the analytic evaluation of these quantities or to provide conflicting indicators.

Example 2.2 Suppose that the prior distribution is altered now from the normal to the Cauchy distribution with the same location μ and scale τ^2 parameters. The posterior density becomes

$$\pi(\theta) \propto \exp\left\{-\frac{1}{2}\frac{(\bar{x}-\theta)^2}{\sigma^2/n}\right\}\frac{1}{\tau^2+(\theta-\mu)^2}$$

and no additional simplification is possible. Unlike the previous example, no functional form can be recognized and none of the summarizing quantities can be obtained analytically. Besides, the form of the posterior can be irregular enough to require extra care when using simple numerical summaries to describe it. Figure 2.1 illustrates a possible form for π where the difficulty in summarizing the information it contains is evident.

The heavy tails of the Cauchy prior are responsible for solving the disagreement between prior and likelihood in favour of the latter in the above example. In general, distributions with thick tails are used to represent such situations where information is not very reliable. Heavy tails distributions were obtained in Chapter 1 as scale mixtures of normal, a device that will be used in the sequel.

2.3 Conjugate distributions

Example 2.1 illustrates a situation where for a given observational model, prior and posterior distributions belong to the same class of distributions, the normal family in that case. This brings advantages to the resulting inferential procedures due to simplification of the analysis that becomes restricted to a subset of all possible distributions. Passage from prior to posterior only involves a change in the hyperparameters with no additional calculation. The distribution of θ can then be routinely updated and the arrival of new observations does not cause any complication.

Example 2.1 (continued) After observing a sample $x_1 = (x_{11}, ..., x_{1n_1})'$ from the $N(\theta, \sigma^2)$ distribution, the posterior for θ is $N(\mu_1, \tau_1^2)$ where $\tau_1^{-2} = n_1\sigma^{-2} + \tau^{-2}$, $\mu_1 = \tau_1^2(n_1\sigma^{-2}\bar{x}_1 + \tau^{-2}\mu)$ and $\bar{x}_1$ is the arithmetic mean for sample x_1. If a new sample $x_2 = (x_{21}, ..., x_{2n_2})'$ from the same population is observed, the posterior for θ given x_1 becomes the prior and the new posterior of interest for θ is conditional on x_1 and x_2. As the prior is normal, derivations above guarantee that the new posterior will also be normal. They also show that the relation between the hyperparameters

(μ_1, τ_1^2) of the new prior and (μ_2, τ_2^2) of the new posterior is

$$\tau_2^{-2} = n_2 \sigma^{-2} + \tau_1^{-2}$$
$$\mu_2 = \tau_2^2 (n_2 \sigma^{-2} \bar{x}_2 + \tau_1^{-2} \mu_1)$$

where $\bar{x}_2$ is the arithmetic mean for sample x_2.

Furthermore, the prior is in general a well known, easy to understand distribution. If the analysis retains the posterior in the same class, the good properties of the distribution will be preserved and the analysis based on the posterior will be simple to summarize.

This preservation of the distribution after updating in the same class defines conjugacy. A family of distributions P is conjugate to an observational model F if for every prior $p \in P$ and for any observational distribution $f \in F$, the posterior $\pi \in P$. Obviously this property is almost always valid for very encompassing classes P, as the class of all continuous distributions, and is seldom valid for very restricted classes, as classes containing a single distribution. The definition will only be useful for classes of reasonable size and is typically used to recognize densities having the same algebraic form apart from a few constants, the hyperparameters. Common sense must be exercised for recognition of useful conjugate classes. A good example is the conjugacy of the normal for normal observations with known variance.

Despite or even because of the advantages of conjugacy, analyses with the conjugate prior must be used with care. The easy calculations of this specification comes with a price due to the restrictions they impose on the form of the prior. In many cases, it is unlikely that the conjugate prior is an adequate representation of one's prior state of uncertainty. The analytic tractability associated with conjugate distributions is lost and posterior summarization becomes harder. This dichotomy between tractability and realism is always present in the applied statistician's work. There is no universal answer for this dilemma and this text aims to provide a solid basis with which model building can be approached with the use of a few examples.

2.3.1 Conjugate distributions for the exponential family

A density or probability function $f(x|\theta)$ belongs to the one-parameter exponential family of distributions if

$$f(x|\theta) = a(x) \exp\{\phi(\theta) t(x) + b(\theta)\} \tag{2.5}$$

The importance of the exponential family is linked to the fact that many of the most used distributions belong to the class. This includes:

a) Normal distribution $N(\theta, \sigma^2)$ (with σ^2 known)

$$\phi(\theta) = \frac{\theta}{\sigma^2}, \ t(x) = x, \ b(\theta) = -\frac{\theta^2}{2\sigma^2}, \ a(x) = \exp\left\{-\frac{x^2}{2\sigma^2}\right\}$$

b) Binomial distribution $bin(n, \theta)$

$$\phi(\theta) = \log\left(\frac{\theta}{1-\theta}\right), \; t(x) = x, \; b(\theta) = n \log(1-\theta), \; a(x) = \binom{n}{x}$$

c) Exponential distribution $Exp(\theta)$

$$\phi(\theta) = -\theta, \; t(x) = x, \; b(\theta) = \log\theta, \; a(x) = 1$$

d) Poisson distribution $Poi(\theta)$

$$\phi(\theta) = \log\theta, \; t(x) = x, \; b(\theta) = -\log\theta, \; a(x) = 1$$

Important families of distributions that do not belong to the exponential family are the uniform, Student's t and discrete mixtures of densities.

The definition (2.5) of the exponential family can be modified in many essentially similar forms. Note that in particular there is no unique form for ϕ and t as multiplication of the first by a constant c can be compensated for by division of the second one by c maintaining the product $\phi(\theta)t(x)$ unchanged. For the exponential distribution, alternative definitions are $\phi(\theta) = \theta$ and $t(x) = -x$.

A simplifying alteration can be obtained by noting that the distribution can be parametrized by $\phi(\theta)$ instead of θ and that it is irrelevant whether $t(x)$ or x was observed. In many cases, in fact, $t(x) = x$. Assuming the existence of the inverse transformations $x(t)$ and $\theta(\phi)$, the density of t is

$$f^*(t|\phi) = a^*(t) \exp\{\phi t + b^*(\phi)\}$$

where $a^*(t) = a(x(t))\left|\frac{dx(t)}{dt}\right|$ and $b^*(\phi) = b(\theta(\phi))$. Both definitions of the exponential family may be used indistinctly once these transformations are recognized. Whenever possible, the second form is preferred in this text for the clear notational simplification. Therefore, a density of the exponential family will from now on be written in the form

$$f(x|\theta) = a(x) \exp\{\theta x + b(\theta)\} \tag{2.6}$$

and θ will be called the canonical parameter. This family has been well studied and

$$E(x|\theta) = \mu = -b'(\theta) \text{ and } Var(x|\theta) = -b''(\theta) \tag{2.7}$$

where b' (b'') denotes the first (second) derivative of the function b, assumed to be a twice differentiable function. The function b'' is sometimes referred to as the variance function (McCullagh and Nelder, 1988). Whenever clear from the context, (2.6) will be denoted by $x \sim EF(\mu)$.

Define now a prior for θ in the form

$$p(\theta) = k(\alpha, \beta) \exp\{\alpha\theta + \beta b(\theta)\}$$

where the normalizing constant k typically depends of the hyperparameters

α and β. Assuming the presence of an observation from the exponential family (2.6) and operating Bayes' theorem gives

$$
\begin{aligned}
\pi(\theta) \quad &\propto \quad f(x|\theta)\,p(\theta) \\
&\propto \quad \exp\{\theta x + b(\theta)\}\,\exp\{\alpha\theta + \beta b(\theta)\} \\
&= \quad \exp\{(\alpha + x)\theta + (\beta + 1)b(\theta)\} \\
&= \quad \exp\{\alpha_1\theta + \beta_1 b(\theta)\} \tag{2.8}
\end{aligned}
$$

where $\alpha_1 = \alpha + x$ and $\beta_1 = \beta + 1$. Therefore, π and p have the same form for model (2.6) and belong to the same class of distributions. Whenever clear from the context, the conjugate prior will be denoted by $\theta \sim CP(\alpha, \beta)$. Equation (2.8) shows that $\theta|x \sim CP(\alpha_1, \beta_1)$.

The derivation above shows that a search of prior distributions with the same algebraic form of the likelihood leads to conjugacy. In specific cases it may be simpler to adopt the strategy of recognition of algebraic forms than specializing the calculation of distributions for canonical parameters that may be harder to handle algebraically.

Example 2.3 Consider the observational model $Poi(\lambda)$ for x and a prior $G(\alpha, \beta)$ for λ. Following Bayes' theorem,

$$
\begin{aligned}
\pi(\lambda) \quad &\propto \quad e^{-\lambda}\lambda^x\,\lambda^{\alpha-1}e^{-\beta\lambda} \\
&\propto \quad \lambda^{\alpha+x-1}e^{-(\beta+1)\lambda}
\end{aligned}
$$

that is the density of a $G(\alpha + x, \beta + 1)$ distribution. Hence, the Gamma family is conjugated to the Poisson observational model. Consideration of the canonical parameter $\log\theta$ is possible but leads to conjugacy of the log Gamma distribution that is more awkward to handle.

There are many other known results about conjugacy inside the exponential family described in Gamerman and Migon (1993). Also, the sequential nature of Bayes' theorem allows that analysis for a sample can be obtained by successive replications where the posterior at the previous step is the prior for the next step. If such an updating operation preserves the family of distributions after one step it generally will after n steps. Therefore, conjugacy with respect to a single observation is equivalent to conjugacy with respect to a sample of size n. At each step of (2.8), updating due to observation x_i through Bayes' theorem reduces to recursively setting hyperparameters $\alpha_i = \alpha_{i-1} + x_i$ and $\beta_i = \beta_{i-1} + 1$, $i = 1, ..., n$ with $\alpha_0 = \alpha$ and $\beta_0 = \beta$. After observing a sample of size n, the posterior and the prior belong to the same class of distributions with $\alpha_n = \alpha + \Sigma_i x_i$ and $\beta_n = \beta + n$ and therefore the model is conjugate. Again, this can be written as $\theta|x_1, ..., x_n \sim CP(\alpha_n, \beta_n)$.

The definition of exponential family can be extended to the multiparameter case. A density or probability function $f(x|\theta)$ belongs to the k-

parameter exponential family of distributions if

$$f(x|\theta) = a(x) \exp\left\{\sum_{i=1}^{k} \theta_i x_i + b(\theta)\right\} \tag{2.9}$$

where $x = (x_1, ..., x_k)'$ and $\theta = (\theta_1, ..., \theta_k)'$.

Example 2.4 A random vector $x = (x_1, ..., x_k)'$ has multinomial distribution with index n and probabilities $p_1, ..., p_k$ satisfying $\Sigma_i x_i = n$ and $\Sigma_i p_i = 1$ if its joint probability function is

$$f(x|p_1, ..., p_k) = \frac{n!}{\prod_{i=1}^{k} x_i!} \prod_{i=1}^{k} p_i^{x_i}, \ 0 \le x_i \le n, \ \Sigma_i x_i = n$$

This probability function can be placed in the form (2.9) with vector $\theta = (\theta_1, ..., \theta_{k-1})'$ of canonical parameters with $\theta_i = log(p_i/p_k)$, $i = 1, ..., k-1$.

A family of conjugate distributions can be constructed in the same way. Define a prior $p(\theta)$ in the form

$$p(\theta) = k(\alpha, \beta) \exp\left\{\sum_{i=1}^{k} \alpha_i \theta_i + \beta b(\theta)\right\}$$

where the normalizing constant k typically depends on hyperparameters $\alpha = (\alpha_1, ..., \alpha_k)'$ and β. Assuming the presence of an observation from the exponential family (2.9), operating Bayes' theorem gives

$$\begin{aligned}
\pi(\theta) \ &\propto \ f(x|\theta)\, p(\theta) \\
&\propto \ \exp\left\{\sum_{i=1}^{k} \theta_i x_i + b(\theta)\right\} \exp\left\{\sum_{i=1}^{k} \alpha_i \theta_i + \beta b(\theta)\right\} \\
&= \ \exp\left\{\sum_{i=1}^{k}(\alpha_i + x_i)\theta_i + (\beta + 1)b(\theta)\right\} \\
&= \ \exp\left\{\sum_{i=1}^{k} \alpha_{1i}\theta_i + \beta_1 b(\theta)\right\}
\end{aligned}$$

where $\alpha_{1i} = \alpha_i + x_i$, $i = 1, ..., k$ and $\beta_1 = \beta + 1$. The densities π and p have the same form for model (2.9) and belong to the same class of distributions and therefore the model is conjugate. Again, in specific cases it may be simpler to adopt the strategy of recognition of algebraic forms than specializing the calculation of distributions for canonical parameters that may be harder to handle algebraically.

Example 2.4 (continued) A random vector $\theta = (\theta_1, ..., \theta_k)'$ subject to the restriction $\Sigma_i \theta_i = 1$ has Dirichlet distribution with parameter $\alpha = (\alpha_1, ..., \alpha_k)$, denoted by $D(\alpha)$, if its density is given by

$$f_D(\theta) = \frac{\Gamma\left(\sum_{i=1}^{k} \alpha_i\right)}{\prod_{i=1}^{k} \Gamma(\alpha_i)} \prod_{i=1}^{k} \theta_i^{\alpha_i - 1}, \quad 0 < \theta_i < 1 \quad (\alpha_i > 0), \; i = 1, ..., k$$

Formally, θ cannot have a density, only its subvectors θ_{-i}, but this mathematical formalism may be dropped in favour of notational simplification.

Combining a prior $D(\alpha)$ for $p = (p_1, ..., p_k)'$ with the multinomial model gives via Bayes' theorem

$$\pi(p_1, ..., p_k) \quad \propto \quad \prod_{i=1}^{k} p_i^{x_i} \prod_{i=1}^{k} p_i^{\alpha_i - 1}$$

$$\propto \quad \prod_{i=1}^{k} p_i^{\alpha_i + x_i - 1}$$

Therefore, $\pi(p) = D(\alpha_1)$ where $\alpha_1 = \alpha + x$ and the family of Dirichlet distributions is conjugate to the multinomial model.

It was shown at the beginning of this section that the normal family of distributions is conjugate to the normal observational model. It is reasonable to consider whether conjugacy is retained when passing to the multivariate case. This result can be established with the help of the properties of the multivariate normal distribution described in section 1.4.

Assume that $\theta \sim N(\mu, B)$ and that $x|\theta \sim N(\theta, \Sigma)$. By the reconstruction of the joint distribution (1.3),

$$\begin{pmatrix} x \\ \theta \end{pmatrix} \sim N\left[\begin{pmatrix} \mu \\ \mu \end{pmatrix}, \begin{pmatrix} \Sigma + B & B \\ B & B \end{pmatrix}\right]$$

The posterior distribution of θ is given by the conditional distribution of $\theta|x$, obtained through (1.2) as

$$\pi(\theta) = N(\mu + B(\Sigma + B)^{-1}(x - \mu), B - B(\Sigma + B)^{-1}B)$$

Therefore also in the multivariate case, the normal model is conjugate to an observational normal model.

2.3.2 Conjugacy and regression models

The results above can be extended to the more general case of normal regression. A response or interest variable is observed and its probabilistic description is known to be affected by other variables, called explanatory variables or covariates. There are many ways these can affect the response. In the simplest case, this influence is linear and additive on the mean response. These assumptions arise as first order approximations for more complicated functions. The influence is taken over the location because it is the most pre-eminent element of a distribution. Other location measures could also have been used.

If the response variables have normal distribution then a linear regression model is obtained with observations $y = (y_1, ..., y_n)'$ described by

$$y_i \sim N(x_{i1}\beta_1 + ... + x_{id}\beta_d, \sigma^2) \quad , \quad i = 1, ..., n$$

where $x_{i1}, ..., x_{id}$ are the values of d explanatory variables for the ith observation and $\beta_1, ..., \beta_d$ are the regression coefficients associated with these variables. For independent observations, the model is written in matrix form as

$$y \sim N(X\beta, \sigma^2 I_n) \tag{2.10}$$

where the $n \times d$ design matrix X and the d-dimensional parametric vector are

$$X = \begin{pmatrix} x_{11} & \cdots & x_{1d} \\ \vdots & & \vdots \\ x_{n1} & \cdots & x_{nd} \end{pmatrix} \text{ and } \beta = \begin{pmatrix} \beta_1 \\ \vdots \\ \beta_d \end{pmatrix}$$

It will be assumed in this text that design matrices are always full rank $d \leq n$. Cases where this does not apply involve redundancies in specification of the model that can in general be removed by simple mathematical operations. In some cases, the homoscedasticity (equal variances) hypothesis is replaced by a more general form where $Var(y) = \text{diag}(\sigma_1^2, ..., \sigma_n^2)$. Even more generally, the generic variance form $Var(y) = \Sigma$ can be used where Σ is any $n \times n$ symmetric positive-definite matrix.

From the Bayesian point of view, the model (2.10) is completed with a prior distribution for parameters (β, σ^2). In this case, there is a conjugate family of distributions given by

$$\beta | \sigma^2 \sim N(b_0, \sigma^2 B_0) \text{ and } \sigma^2 \sim IG\left(\frac{n_0}{2}, \frac{n_0 S_0}{2}\right)$$

The unconventional form for the prior hyperparameters (especially from the distribution of σ^2) will be useful below but note that the above is equivalent to $n_0 S_0 / \sigma^2 \sim \chi_{n_0}^2$. Before establishing this conjugacy, another reparametrization will be adopted: σ^2 will be replaced by the parameter $\phi = \sigma^{-2}$. In terms of ϕ, the prior is $NG(b_0, B_0, n_0, S_0)$, that is,

$$\beta | \phi \sim N(b_0, \phi^{-1} B_0) \text{ and } \phi \sim G\left(\frac{n_0}{2}, \frac{n_0 S_0}{2}\right)$$

(and hence $n_0 S_0 \phi \sim \chi_{n_0}^2$) and its density is

$$p(\beta, \phi) \propto p(\beta | \phi) p(\phi)$$

$$\propto \quad \phi^{d/2} \exp\left\{-\frac{\phi}{2}(\beta - b_0)'B_0^{-1}(\beta - b_0)\right\} \phi^{(n_0/2)-1} \exp\left\{-\frac{\phi}{2}n_0 S_0\right\}$$

$$\propto \quad \phi^{[(n_0+d)/2]-1} \exp\left\{-\frac{\phi}{2}[n_0 S_0 + (\beta - b_0)'B_0^{-1}(\beta - b_0)]\right\} \tag{2.11}$$

The likelihood (2.10) can also be written in terms of β and ϕ as $y|\beta, \phi \sim N(X\beta, \phi^{-1}I_n)$ and

$$
\begin{aligned}
l(\beta, \phi) &= f(y|\beta, \phi) \\
&\propto \phi^{-n/2} \exp\left\{-\frac{\phi}{2}(y - X\beta)'(y - X\beta)\right\} \\
&= \phi^{-n/2} \exp\left\{-\frac{\phi}{2}[Q(\beta) + S_e]\right\} \tag{2.12}
\end{aligned}
$$

where $Q(\beta) = (\beta - \hat{\beta})'X'X(\beta - \hat{\beta})$, $\hat{\beta} = (X'X)^{-1}X'y$ and $S_e = (y - X\hat{\beta})'(y - X\hat{\beta}) = \sum_{i=1}^{n} e_i^2$ with $e_i = y_i - \hat{\mu}_i$ and $\hat{\mu}_i = x_{i1}\hat{\beta}_1 + ... + x_{id}\hat{\beta}_d$, $i = 1, ..., n$. Note from (2.12) that $\hat{\beta}$ maximizes $l(\beta, \phi)$ as a function of β.

Combining (2.11) and (2.12) gives the posterior

$$
\begin{aligned}
\pi(\beta, \phi) &\propto l(\beta, \phi) p(\beta, \phi) \\
&\propto \phi^{\frac{n_0+n+d}{2}-1} \exp\left\{-\frac{\phi}{2}[n_0 S_0 + S_e + (\beta - b_0)'B_0^{-1}(\beta - b_0) + Q(\beta)]\right\} \\
&\propto \phi^{\frac{d}{2}} \exp\left\{-\frac{\phi}{2}(\beta - b_1)'B_1^{-1}(\beta - b_1)\right\} \phi^{\frac{n_1}{2}-1} \exp\left\{-\frac{\phi}{2}n_1 S_1\right\} \tag{2.13}
\end{aligned}
$$

where $n_1 = n_0 + n$, $n_1 S_1 = n_0 S_0 + (y - X b_1)'y + (b_0 - b_1)'B_0^{-1}b_0$, $b_1 = B_1(B_0^{-1}b_0 + X'y)$ and $B_1^{-1} = B_0^{-1} + X'X$. Comparing (2.11) and (2.13) it is clear that posterior and prior have the same form and belong to a conjugate family of distributions. The posterior for (β, ϕ) is $NG(b_1, B_1, n_1, S_1)$. If the posterior is required in terms of σ^2, the same transformation operated over the prior can be applied and

$$\beta|\sigma^2 \sim N(b_1, \sigma^2 B_1) \text{ and } \sigma^2 \sim IG\left(\frac{n_1}{2}, \frac{n_1 S_1}{2}\right)$$

Another form to rewrite the normal regression form in is

$$y_i \sim N(\mu_i, \sigma^2) \text{ with } \mu_i = x_{i1}\beta_1 + ... + x_{id}\beta_d$$

for $i = 1, ..., n$. This form is particularly useful for extensions of regression models to other observational distributions. The linear predictor defined by $\eta_i = x_{i1}\beta_1 + ... + x_{id}\beta_d$ brings into the model the joint effect of the explanatory variables. It is typically a real number as no restrictions are imposed on the values of the covariates. The normal mean μ_i can also be any real number and there is no problem in equating the two parameters. The same does not necessarily hold for other distributions as will be seen below.

An extension of regression models still preserving linearity and influence of covariates through the mean response is given by generalized linear models. The observations remain independent but now have distributions in the exponential family. The model is

$$
\begin{aligned}
f(y_i|\theta_i) &= a(y_i)\exp\{y_i\theta_i + b(\theta_i)\} \text{ with } E(y_i|\theta_i) = -b'(\theta_i) = \mu_i\\
g(\mu_i) &= \eta_i\\
\eta_i &= x_{i1}\beta_1 + \ldots + x_{id}\beta_d
\end{aligned}
$$

for $i = 1, ..., n$, where the link function g is differentiable.

Example 2.5 Consider $y_i|\pi_i \sim bin(n_i, \pi_i)$, $i = 1, ..., n$, and assume that the probabilities π_i are determined by the values of a variable x. The π_i lie between 0 and 1 and can be associated to a distribution function. One possibility is the normal distribution and in this case

$$
\pi_i = \Phi(\alpha + \beta x_i) \quad , \quad i = 1, ..., n
$$

where Φ is the d.f. of the $N(0,1)$ distribution and α and β are constants. The binomial distribution belongs to the exponential family and the link function $g_1 = \Phi^{-1}$ is differentiable. The structure of a generalized linear model is completed with the linear predictor

$$
\eta_i = \alpha + \beta x_i \quad , \quad i = 1, ..., n
$$

Other possible links include the logistic and complementary log-log transformations

$$
g_2(\pi_i) = \text{logit}(\pi_i) = \log\left(\frac{\pi_i}{1 - \pi_i}\right) \text{ and } g_3(\pi_i) = \log[-\log(1 - \pi_i)]
$$

associated respectively to the logistic and extreme-value distributions. Note that g_1, g_2 and g_3 take numbers from $[0,1]$ to the real line.

Example 2.6 Consider $y_i|\lambda_i \sim Poi(\lambda_i)$, $i = 1, ..., n$, containing counts of events affected by values of variables $x_1, \ldots, x_d$. As $E(y_i|\lambda_i) = \lambda_i > 0$, it is not recommended to equate λ_i to the linear predictor η_i that can take any value in the line. A simple solution is to take the $\log(\lambda_i)$ transformation. The effects of the variables $x_1, \ldots, x_p$ become multiplicative on the mean as

$$
\lambda_i = \exp(x_{i1}\beta_1 + \ldots + x_{id}\beta_d)
$$

The theory of generalized linear models as well as many applications, special cases and maximum likelihood estimation are very well described in McCullagh and Nelder (1988). For Bayesian inference, a prior distribution for the parametric vector β must be specified. A natural extension of the normal regression is to take a $N(b_0, B_0)$ prior. Unfortunately, this distribution is not conjugate and exact Bayesian inference is not possible. The same comment applies to other prior distributions with rare and not useful exceptions. This is an example of an important class of models from the practical point of view where inference procedures cannot be exactly obtained.

2.3.3 Conditional conjugacy

The difficulty in obtaining conjugate distributions increases with the complexity of the model and the dimension of the parametric space. It was mentioned in the previous subsection that the important class of generalized linear models does not admit analytically tractable conjugacy. In other instances, the model is structured in such a way that some elements of the parametric vector have conjugate distributions but the vector as a whole does not possess conjugacy.

Consider again the normal regression model (2.10) and assume now that β and ϕ are prior independent with distributions $\beta \sim N(b_0, B_0)$ and $\phi \sim G(n_0/2, n_0 S_0/2)$. The change with respect to the previous specification is the removal of the dependence between β and ϕ in the conditional variance of $\beta|\phi$. The posterior distribution now has density

$$
\begin{aligned}
\pi(\beta, \phi) \;\propto\;& l(\beta, \phi)\, p(\beta)\, p(\phi) \\
\propto\;& \phi^{n/2} \exp\left\{ -\frac{\phi}{2}[(\beta - \hat{\beta})' X' X (\beta - \hat{\beta}) + S_e] \right\} \\
\times\;& \exp\left\{ -\frac{1}{2}[(\beta - b_0)' B_0^{-1}(\beta - b_0)] \right\} \phi^{n_0/2-1} \exp\left\{ -\frac{n_0 S_0 \phi}{2} \right\} \\
\propto\;& \phi^{(n+n_0)/2-1} \exp\left\{ -\frac{\phi}{2}[(\beta - \hat{\beta})' X' X (\beta - \hat{\beta}) + S_e + n_0 S_0] \right\} \\
\times\;& \exp\left\{ -\frac{1}{2}[(\beta - b_0)' B_0^{-1}(\beta - b_0)] \right\} \quad (2.14)
\end{aligned}
$$

and it is not possible to recognize any analytically tractable form as the prior. Therefore, there is no conjugacy. Formally, it would still be possible to establish conjugacy because the prior is a special case of the posterior. This line of thought has little practical value. The notion of conjugacy is only useful when it leads to a form that can be treated analytically, which is not the case here.

Despite the lack of conjugacy, it is easy to obtain the conditional posterior distributions of $\beta|\phi$ and $\phi|\beta$. The first one is

$$
\begin{aligned}
\pi(\beta|\phi) \;\propto\;& \exp\left\{ -\frac{1}{2}[\phi(\beta - \hat{\beta})' X' X (\beta - \hat{\beta}) + (\beta - b_0)' B_0^{-1}(\beta - b_0)] \right\} \\
\propto\;& \exp\left\{ -\frac{1}{2}[(\beta - b_\phi)' B_\phi^{-1}(\beta - b_\phi)] \right\} \quad (2.15)
\end{aligned}
$$

where $b_\phi = B_\phi(B_0^{-1}b_0 + \phi X'y)$ and $B_\phi^{-1} = B_0^{-1} + \phi X'X$. Hence, $\beta|\phi \sim N(b_\phi, B_\phi)$ a posteriori. It has the same form of the conditional prior $\beta|\phi \sim N(b_0, B_0)$, by the independence between β and ϕ. So, β is conditionally conjugate.

Similarly, the conditional posterior distribution of $\phi|\beta$ is

$$\pi(\phi|\beta) \propto \phi^{(n+n_0)/2-1} \exp\left\{-\frac{\phi}{2}[(\beta - \hat{\beta})'X'X(\beta - \hat{\beta}) + S_e + n_0 S_0]\right\}$$

and therefore, a posteriori $\phi|\beta \sim G(n_1/2, n_1 S_\beta/2)$ where $n_1 = n + n_0$ and $S_\beta = (\beta - \hat{\beta})'X'X(\beta - \hat{\beta}) + S_e + n_0 S_0$. Again, this conditional posterior has the same form of the conditional prior. So, there is conjugacy for both conditional distributions.

The above specification is of practical relevance as it is conceivable that one would want to set initial uncertainty through independent forms for β and ϕ. The conditional priors simplify to the marginal priors. Nevertheless, it is possible to have conditional conjugacy even when the priors are not independent (Exercise 2.9).

In the case of a parametric vector $\theta = (\theta_1, ..., \theta_d)'$, a (not necessarily scalar) component θ_i, $i = 1, ..., d$, exhibits conditional conjugacy if the full conditional prior $p_i(\theta_i) = p(\theta_i|\theta_{-i})$ and the full conditional posterior $\pi_i(\theta_i) = \pi(\theta_i|\theta_{-i})$ belong to the same family of distributions. Depending on the form of the prior, it is possible that all components of θ are conditionally conjugate. In other cases, only a few components will be conditionally conjugate (see Exercise 2.9c).

Conditional conjugacy is present in many complex models where complete probabilistic specification is only possible when the parametric space is qualitatively structured before prior quantification takes place. In highly dimensional parametric space it is very hard to directly specify a joint prior distribution. It makes sense to explore qualitative relations of (conditional) independence between the components of the parameter vector. In many cases, this will lead to substantial simplifications in the form of the prior. It will be formed by a collection of probabilistic statements about components of smaller dimensions. These are easier to specify and quantify. Important examples of this qualitative approach to model building are provided in the next sections. Due to the impossibility of exact analysis, inference procedures exploring full conditional distributions and, whenever possible, conditional conjugacy becomes of great importance.

2.4 Hierarchical models

The normal regression model was specified in the previous section with the aim of establishing relations between the response variable y and a set of explanatory variables $x_1, \ldots, x_p$ through regression coefficients gathered in the vector β. Many times, the problem is structured in a way that qualitative probabilistic statements about β can and should be incorporated into the model.

Example 2.7 Consider observations $y_{ij} \sim N(\beta_i, \sigma^2)$, $j = 1, ..., n_i$, $i = 1, ..., d$, collected from d groups with different means β_i but the same dispersion. This model is a special case of a regression model with observation vector $y = (y_{11}, ..., y_{1n_1}, ..., y_{d1}, ..., y_{dn_d})'$ and design matrix $X = \text{diag}(1_{n_1}, ..., 1_{n_d})$ where 1_m is the m-dimensional vector of 1s. The model is completed with a prior distribution for (β, σ^2). One possibility is to assume prior independence between the means β_i, $i = 1, ..., d$. If the d groups are similar in some sense, a plausible alternative is to assume that the means are a sample from a population of means. This population may be hypothetical and, to fix ideas, is assumed here to be homogeneous. Assuming a normal population, $\beta_1, ..., \beta_d$ is a sample from a $N(\mu, \tau^2)$ where μ is the mean and τ^2 measures the dispersion of the population of means. The model is completed with a prior distribution for (μ, τ^2). The complete prior specification is

$$
\begin{aligned}
\text{1st level} \quad &: \quad \beta|\mu, \tau^2 \sim N(1_d\mu, \tau^2 I_d) \\
\text{2nd level} \quad &: \quad \mu \sim N(b_0, B_0) \\
\sigma^2 \sim F_\sigma \quad &\text{and} \quad \tau^2 \sim F_\tau
\end{aligned}
$$

for independent probability distributions F_σ and F_τ. The prior density for the model parameters $(\beta, \mu, \sigma^2, \tau^2)$ is

$$
p(\beta, \mu, \sigma^2, \tau^2) = p(\beta|\mu, \tau^2)\, p(\mu) p(\sigma^2) p(\tau^2)
$$

Note that the prior in the example was specified in two stages. The (two-stage) model of the example can be generalized in many ways. A generalization towards a normal regression model was considered by Lindley and Smith (1972) and is given by

$$
\begin{aligned}
y \mid \beta_1, \phi \quad &\sim \quad N(X_1\beta_1, \phi^{-1}I_n) \\
\beta_1 \mid \beta_2 \quad &\sim \quad N(X_2\beta_2, C) \\
\beta_2 \quad &\sim \quad N(b, B) \\
\phi \quad &\sim \quad G(n_0/2, n_0\sigma_0^2/2)
\end{aligned} \tag{2.16}
$$

The design matrix with the covariates for the response vector y and the regression coefficient were respectively renamed to X_1 and β_1. This is due to the presence of another design matrix containing a further set of explanatory variables X_2 and another regression coefficient β_2. This matrix contains the values that explain the variations of the values of β_1, and β_2 contains the coefficients of this explanation.

The prior in the above model is specified in two stages. Depending on the problem, more stages may be required for an appropriate description of the model. Its form may remain unchanged with additional equations in the form

$$\beta_j \mid \beta_{j+1} \sim N(X_{j+1}\beta_{j+1}, C_j)$$

In general, the higher the stage, the harder is the specification of the distributions. Rarely, models have more than three stages and it is very common that the prior at the higher stage is set to be non-informative.

So far, nothing has been said about the matrices C and B being assumed to be known. This assumption is not reasonable in general and a modification sometimes suggested is the substitution of C and B by $\phi^{-1}C$ and $\phi^{-1}B$ respectively. The variances C and B will then measure prior dispersion relatively to the likelihood dispersion. This dependence on ϕ allows the use of the results about conjugacy. The analysis still remains conditional on the (assumed known) values of the matrices C and B. This procedure was adopted by Gamerman and Migon (1993, Chapter 7) but will not be pursued here. Details of the relevant distributions are left as an exercise.

The derivations below concentrate on the two-stage model to simplify the notation even though there is no technical problem in the extension to the k-stage models, $k > 2$. The first point to mention is that the structure imposed upon the joint distribution of all the variables in the problem, that is $(y, \beta_1, \beta_2, \phi)$, is

$$p(y, \beta_1, \beta_2, \phi) = p(y|\beta_1, \phi)p(\beta_1|\beta_2)p(\beta_2)p(\phi)$$

The hierarchical character of the model then becomes clear with the successive conditional specifications. Unfortunately, the analysis is not analytically tractable and it is not possible to obtain the marginal posterior distributions of β_1 and ϕ. The analysis conditional on knowledge of β_2 is not new and was performed in the previous section. If β_2 is known, the prior does not depend on the probabilistic specification of β_2. Replacement of b_0 by $X_2\beta_2$ in the regression model is all that needs to be done. Hence, the full conditional posterior for β_1 is $N(b_\phi, B_\phi)$ and for ϕ is $G(n_1/2, n_1S_\beta/2)$ where $b_\phi = B_\phi(C^{-1}X_2\beta_2 + \phi X_1'y)$, $B_\phi = C^{-1} + \phi X_1'X_1$, $n_1 = n + n_0$ and $n_1S_\beta = n_0\sigma_0^2 + (y - X_1\beta_1)'(y - X_1\beta_1)$. If β_2 is unknown, it is also not possible to obtain its marginal distribution in closed form but its full conditional posterior distribution is

$$
\begin{aligned}
\pi(\beta_2|\beta_1, \phi) \quad &\propto \quad p(\beta_1|\beta_2)p(\beta_2) \\
&\propto \quad \exp\left\{-\frac{1}{2}(\beta_1 - X_2\beta_2)'C^{-1}(\beta_1 - X_2\beta_2)\right\} \\
&\times \quad \exp\left\{-\frac{1}{2}(\beta_2 - b)'B^{-1}(\beta_2 - b)\right\} \\
&\propto \quad \exp\left\{-\frac{1}{2}(\beta_2 - b^*)'B^{*-1}(\beta_2 - b^*)\right\}
\end{aligned}
$$

where $b^* = B^*(B^{-1}b + X_2'C^{-1}\beta_1)$ and $B^* = (B^{-1} + X_2'C^{-1}X_2)^{-1}$. Therefore, all model parameters β_1, β_2 and ϕ are conditionally conjugate.

It is interesting to note that the full conditional posterior distribution of β_2 does not depend on the observation. This somewhat surprising fact is a direct consequence of the hierarchical structure of model that passes through β_1 all information provided by y to β_2. More formally, y and β_2 are conditionally independent given β_1. This model building strategy based on conditional independence allows easy derivation of full conditional distributions. In many cases, they are also conditionally conjugate.

There is nothing special about the normal distribution with respect to hierarchical models. They can also be defined for other observational distributions. Example 2.7 can be extended for observations y_{ij} with density $f(y|\beta_i, \phi)$. The β_i are a sample from a population with density $p(\beta|\lambda)$ and this constitutes the first stage of the prior distribution. Again, the model is completed with a second stage prior for λ and ϕ (Deely and Lindley, 1981).

Kass and Steffey (1989) study this class of models and call them conditionally independent hierarchical models. A case of particular interest is when the βs are conditionally conjugate given the values of ϕ and λ. In the context of exponential family distributions, $y_{ij} \sim EF(\mu_i)$, the θ_i form a sample from a $CP(\alpha, \beta)$ distribution and the model is completed with a prior for α and β. Then, the full conditional posterior of the θ_i is still given by a product of independent $CP(\alpha_i^*, \beta_i^*)$ distributions. George, Makov and Smith (1993) show that the full conditional posterior distributions of α and β are log-concave. Hence, despite their awkward functional form (Exercise 2.13) they are available for adaptive rejection sampling.

Albert (1996) discusses the possibility of using other non-conjugate specifications for the first stage prior. In particular, he considers parametric transformations taking the β_i to the real line and suggests a normal first stage prior to the transformed parameter as done in Example 2.7.

A more general extension of the hierarchical model is considered in Albert (1988). The two-stage regression model (2.16) and generalized linear models can be combined to give a generalized linear hierarchical model

$$
\begin{aligned}
y_i \mid \mu_i &\sim EF(\mu_i) \quad, \quad i = 1, ..., n \\
\eta &= X_1\beta_1 \\
\beta_1 \mid \beta_2 &\sim N(X_2\beta_2, C) \\
\beta_2 &\sim N(b, B)
\end{aligned}
\tag{2.17}
$$

where $\eta = (\eta_1, ..., \eta_n)'$ and $\eta_i = g(\mu_i)$, $i = 1, ..., n$. When the EF distribution considered is the normal, model (2.16) is obtained. When the regression structure simply classifies observations by groups, the normal prior model for transformed means suggested by Albert (1996) is obtained. Again, there is nothing preventing the inclusion of further stages in the hierarchy although this is seldom required.

2.5 Dynamic models

A large class of models with time-varying parameters, adequate to the modelling of time series and regression, was presented by Harrison and Stevens (1976) and is well described in the book by West and Harrison (1997). This section present some basic aspects of dynamic models with a few illustrative examples.

Dynamic linear models are defined by a pair of equations, called the observation equation and the evolution or system equation, respectively given by

$$y_t = F_t'\beta_t + \epsilon_t, \qquad \epsilon_t \sim N(0, \sigma_t^2) \tag{2.18}$$
$$\beta_t = G_t\beta_{t-1} + \omega_t, \qquad \omega_t \sim N(0, W_t) \tag{2.19}$$

where $\{y_t\}$ is a sequence of observations through time, conditionally independent given β_t and σ_t^2, F_t is a vector of explanatory variables as in the previous sections, β_t is a d-dimensional vector of regression coefficients or state parameters at time t and G_t is a $d \times d$ matrix describing the parametric evolution. The errors ϵ_t and w_t are mutually independent and σ_t^2 and W_t are the error variances respectively associated to the univariate observation and the d-dimensional vector of parameters. The model is completed with a prior $\beta_1 \sim N(a, R)$.

Dynamic linear models provide another nice example of specification of a prior for a highly dimensional parameter by combination of qualitative and quantitative information. The system equation provides qualitative information about the relation between successive values of the state parameters. Quantitative information is provided by the prior distributions of β_1 and evolution errors w_t. The complete expression of the prior distribution results from the combination of these sources of information.

Dynamic regression models are defined by $G_t = I_d, \forall t$. If, in addition, $W_t = 0, \forall t$, the static regression model (2.10) is obtained. This is equivalent to setting the regression coefficients β_t fixed in time.

Example 2.8 The simplest time series model is the first order model. It is given by equations

$$y_t = \beta_t + \epsilon_t, \qquad \epsilon_t \sim N(0, \sigma_t^2)$$
$$\beta_t = \beta_{t-1} + \omega_t, \qquad \omega_t \sim N(0, W_t)$$

and β_t is scalar. The model can be thought of as a first order Taylor series approximation of a smooth function representing the time trend of the series. This model is useful for stock control, production planning and financial data analysis. Observational and system variances may evolve in time, offering great scope for modelling the variability of the system.

The linear growth model is slightly more elaborate by incorporation of an extra time-varying parameter β_2 representing the growth of the level of the series. The model becomes

$$
\begin{aligned}
y_t &= \beta_{1,t} + \epsilon_t \quad \epsilon_t \sim N(0, \sigma_t^2) \\
\beta_{1,t} &= \beta_{1,t-1} + \beta_{2,t} + \omega_{1,t} \\
\beta_{2,t} &= \beta_{2,t-1} + \omega_{2,t} \quad , \quad \omega_t = (\omega_{1,t}, \omega_{2,t})' \sim N(0, W_t)
\end{aligned}
$$

This model can be written in the form (2.18)-(2.19) with $F_t = (1,0)$ and
$G_t = \begin{pmatrix} 1 & 1 \\ 0 & 1 \end{pmatrix}, \forall t.$

The choice of F_t and G_t depends on the desired model and the nature of the series one wishes to describe. Complete specification of the model requires full description of the variances σ_t^2 and W_t. In general they are assumed to be constant in time with σ_t^2 typically larger than the entries of W_t in applications.

Exact inference is only possible when W_t is replaced by $\sigma_t^2 W_t$. The matrix W_t becomes a matrix of weights relative to the observational variance. Typically it is unknown and must be estimated. This will make analytical treatment impossible. To avoid it, West and Harrison (1997) suggested it should be subjectively assessed through the concept of discount factors. Here, none of these simplifying assumptions are made in preparation for a full analysis described in later chapters.

Nevertheless, consider initially that σ_t^2 and W_t are known and let $y^t = \{y_t, y^{t-1}\}$ with y^0 describing the initial information available, including the values of F_t and G_t, $\forall t$ also assumed known here. Observations are independent conditionally on the state parameters. This time indexed structure matches well with the Bayesian approach by the easy accommodation of sequential procedures and subjective specifications.

2.5.1 Sequential inference

One of the main aspects of a dynamic model is that at any time t, inference can be based on the updated distribution of $\beta_t | y^t$. Sequential inference then carries this through time. There are three basic operations involved here: evolution, prediction and updating. These operations are presented here in this order.

Consider that at time $t - 1$, the updated distribution is $\beta_{t-1} | y^{t-1} \sim N(m_{t-1}, C_{t-1})$. The system equation can be written as $\beta_t | \beta_{t-1} \sim N(G_t \beta_{t-1}, W_t)$. These specifications can be combined according to (1.3) leading to the marginal distribution

$$
\beta_t | y^{t-1} \sim N(a_t, R_t) \tag{2.20}
$$

with $a_t = G_t m_{t-1}$ and $R_t = G_t C_{t-1} G_t' + W_t$;

One-step-ahead prediction can be made by noting that

$$p(y_t, \beta_t | y^{t-1}) = p(y_t | \beta_t) p(\beta_t | y^{t-1})$$

Again, the joint distribution of $y_t, \beta_t | y^{t-1}$ can be reconstructed using (1.3) and the marginal distribution

$$y_t | y^{t-1} \sim N(f_t, Q_t) \qquad (2.21)$$

with $f_t = F_t' a_t$ and $Q_t = F_t' R_t F_t + \sigma_t^2$ is obtained.

Finally, updating is achieved by the standard Bayes' theorem operation of including the observed y_t into the set of available information. The updated posterior distribution is obtained by

$$p(\beta_t | y^t) = p(\beta_t | y_t, y^{t-1}) \propto p(y_t | \beta_t) \, p(\beta_t | y^{t-1})$$

The resulting posterior distribution is

$$\beta_t | y^t \sim N(m_t, C_t) \qquad (2.22)$$

with $m_t = a_t + A_t e_t$ and $C_t = R_t - A_t A_t' Q_t$, where $A_t = R_t F_t / Q_t$ and $e_t = y_t - f_t$. This result follows from (2.15) with prior (2.20) and likelihood (2.18) and from the identity $C_t^{-1} = R_t^{-1} + F_t' F_t \sigma_t^{-2}$. It is sometimes referred to as the Kalman filter.

By induction, these distributions are valid for all times as β_1 also has prior in the form (2.20) and therefore, its posterior must be given by (2.22) with $t = 1$.

2.5.2 Smoothing

The joint distribution of $y^n = (y_1, ..., y_n)'$ and $\beta = (\beta_1, ..., \beta_n)'$ has density

$$p(y^n, \beta) = \prod_{t=1}^{n} p(y_t | \beta_t) \prod_{t=2}^{n} p(\beta_t | \beta_{t-1}) \, p(\beta_1)$$

Therefore, the full conditional density[†] of β_t is

$$
\begin{aligned}
\pi_t(\beta_t) &\propto p(y_t | \beta_t) \, p(\beta_{t+1} | \beta_t) \, p(\beta_t | \beta_{t-1}) \\
&\propto f_N(y_t; F_t' \beta_t, \sigma_t^2) \, f_N(\beta_{t+1}; G_{t+1} \beta_t, W_{t+1}) \, f_N(\beta_t; G_t \beta_{t-1}, W_t) \\
&= f_N(\beta_t; b_t, B_t) \qquad (2.23)
\end{aligned}
$$

where $b_t = B_t(\sigma_t^2 F_t y_t + G_{t+1}' W_{t+1}^{-1} \beta_{t+1} + W_t^{-1} G_t \beta_{t-1})$ and $B_t = (\sigma_t^2 F_t F_t' + G_{t+1}' W_{t+1}^{-1} G_{t+1} + W_t^{-1})^{-1}$ for $t = 2, ..., n-1$. The endpoint parameters β_1 and β_n also have full conditional distributions $N(b_1, B_1)$ and $N(b_n, B_n)$ where $b_1 = B_1(\sigma_1^2 F_1 y_1 + G_2' W_2^{-1} \beta_2 + R^{-1} a)$, $B_1 = (\sigma_1^2 F_1 F_1' + G_2' W_2^{-1} G_2 + R^{-1})^{-1}$, $b_n = B_n(\sigma_n^2 F_n y_n + W_n^{-1} G_n \beta_{n-1})$ and $B_n = (\sigma_n^2 F_n F_n' + W_n^{-1})^{-1}$.

[†] The use of the symbol π for a density automatically implies it was obtained conditional on all available data information, in this case, y^n.

The distribution of parameters at time t can be revised after data at times subsequent to t becomes available. A set of distributions of $p(\beta_t|y^{t+k})$, for k integer, can be considered. When $k > 0$, they are called smoothed or filtered distributions of the parameters. When $k = 0$, it is the updated distribution and when $k < 0$, they are prior distributions. In dynamic models, the smoothed distribution $\pi(\beta|y^n)$ is more commonly used. It has density

$$\pi(\beta|y^n) = p(\beta_n|y^n) \prod_{t=1}^{n-1} p(\beta_t|\beta_{t+1}, ..., \beta_n, y^n)$$

$$= p(\beta_n|y^n) \prod_{t=1}^{n-1} p(\beta_t|\beta_{t+1}, y^t) \tag{2.24}$$

where the last equality follows from the fact that given β_{t+1}, β_t is independent of all quantities indexed by times larger than t. Integrating (2.24) with respect to $(\beta_1, ..., \beta_{t-1})$ gives

$$\pi(\beta_t, ..., \beta_n|y^n) = p(\beta_n|y^n) \prod_{k=t}^{n-1} p(\beta_k|\beta_{k+1}, y^t)$$

for $t = 1, ..., n-1$ and

$$\pi(\beta_t, \beta_{t+1}|y^n) = p(\beta_{t+1}|y^n)p(\beta_t|\beta_{t+1}, y^t) \tag{2.25}$$

for $t = 1, ..., n-1$.

Equation (2.25) provides a simple and recursive form to obtain the marginal posterior distributions of $\beta_t|y^n$. After sequentially obtaining the updated distributions of $\beta_t|y^t$ for $t = 1, ..., n$, time orientation is reversed from the distribution of $\beta_n|y^n$ so as to sucessively obtain the distributions of $\beta_t|y^n$ for $t = n-1, ..., 1$. It can be shown (Exercise 2.18) that

$$\beta_t|y^n \sim N(m_t^n, C_t^n) \tag{2.26}$$

where

$$m_t^n = m_t + C_t G'_{t+1} R_{t+1}^{-1}(m_{t+1}^n - a_{t+1})$$
$$C_t^n = C_t - C_t G'_{t+1} R_{t+1}^{-1}(R_{t+1} - C_{t+1}^n) R_{t+1}^{-1} G_{t+1} C_t$$

Other important aspects of dynamic models are the treatment of missing observations and interventions. In the first case, updating equations are simply not used or the corresponding term $p(y_t|\beta_t)$ not included in the calculations. Interventions are accommodated by use of other distributions for the disturbances w_t with increased uncertainty for times of structural changes. Recent observations will be more heavily weighted when used in the updating step. If the disturbances remain normally distributed, minor changes are required. Scale mixtures of normal are again an alternative worth considering but exact inferences will no longer be possible.

2.5.3 Extensions

Returning to the case of σ_t^2 and W_t all unknown, all the above results remain valid conditionally on the σ_t^2s and W_ts. Equation (2.23) provides the full conditional posterior of β_t and equation (2.24) provides the full conditional posterior of β.

Assume now that $\sigma_t^2 = \sigma^2 = \phi^{-1}$ and $W_t = W = \Phi^{-1}$, $\forall t$. It is not possible to obtain analytic expressions for the posterior marginal densities of β, ϕ and Φ. Nevertheless, it is easy to obtain the full conditional densities of ϕ (or σ^2) and Φ (or W)

$$\pi(\phi|\beta, \Phi) \quad \propto \quad \prod_{t=1}^{n} p(y_t|\beta_t, \phi)\, p(\phi|\beta, \Phi)$$

$$\pi(\Phi|\beta, \phi) \quad \propto \quad \prod_{t=2}^{n} p(\beta_t|\beta_{t-1}, \Phi)\, p(\Phi|\beta, \phi)$$

If, a priori, $\phi \sim G(n_\sigma/2, n_\sigma S_\sigma/2)$ and $\Phi \sim W(n_W/2, n_W S_W/2)$ are independent then

$$\pi(\phi|\beta, \Phi) \quad \propto \quad \prod_{t=1}^{n} f_N(y_t; F_t'\beta_t, \phi^{-1})\, f_G(\phi; n_\sigma/2, n_\sigma S_\sigma/2)$$

$$\propto \quad f_G(\phi; n_\sigma^*/2, n_\sigma^* S_\sigma^*/2) \tag{2.27}$$

$$\pi(\Phi|\beta, \phi) \quad \propto \quad \prod_{t=2}^{n} f_N(\beta_t; G_t\beta_{t-1}, \Phi^{-1})\, f_W(\Phi; n_W/2, n_W S_W/2)$$

$$\propto \quad f_W(\Phi; n_W^*/2, n_W^* S_W^*/2) \tag{2.28}$$

where $n_\sigma^* = n_\sigma + n$, $n_\sigma^* S_\sigma^* = n_\sigma S_\sigma + \Sigma_t(y_t - F_t'\beta_t)^2$, $n_W^* = n_W + n - 1$ and $n_W^* S_W^* = n_W S_W + \Sigma_{t=2}^{n}(\beta_t - G_t\beta_{t-1})(\beta_t - G_t\beta_{t-1})'$. Therefore, a posteriori, $(\sigma^2|\beta, W) \sim IG(n_\sigma^*/2, n_\sigma^* S_\sigma^*/2)$ and $(W|\beta, \sigma^2) \sim IW(n_W^*/2, n_W^* S_W^*/2)$. ϕ and Φ (or equivalently σ^2 and W) exhibit conditional conjugacy.

An extension of dynamic models still preserving linearity and influence through mean response is given by generalized linear models where the observations now have distribution in the exponential family. The model is

$$
\begin{aligned}
f(y_t|\theta_t) &= a(y_t)\, \exp\{y_t\theta_t + b(\theta_t)\} \text{ with } E(y_t|\theta_t) = \mu_t \\
g(\mu_t) &= F_t'\beta_t \\
\beta_t &= G_t\beta_{t+1} + w_t \text{ with } w_t \sim N(0, W_t)
\end{aligned}
\tag{2.29}
$$

where the link function g is again differentiable. The model is completed with a prior $\beta_1 \sim N(a, R)$. It combines the prior specification of normal dynamic models with the observational structure of generalized linear models.

Examples include the dynamic logistic regression with a series of binomial observations y_t with respective success probabilities π_t dynamically related to explanatory variables $x = (x_1, ..., x_d)'$ through the logistic link logit$(\pi_t) = x_t'\beta_t$, and series of Poisson counts with means λ_t dynamically related through multiplicative perturbations $\lambda_t = \lambda_{t-1}w_t^*$. After a logarithmic transformation, one obtains $\log \lambda_t = \log \lambda_{t-1} + w_t$ with $w_t = \log w_t^*$ which is in the form (2.19).

Dynamic generalized models were introduced by West, Harrison and Migon (1985). As with generalized linear models, it is no longer possible to perform exact Bayesian inference and they proposed the use of approximations coupled with a conjugate analysis. Again, an important class of models of practical relevance is obtained for which inference cannot be performed exactly.

2.6 Exercises

1. Show that the posterior distribution of θ, $p(\theta|x, y)$, does not depend on the order in which x and y were processed, that is, it is the same whether one obtains the posterior of $\theta|x$ and uses this posterior as the prior for observation y, one obtains the posterior of $\theta|y$ and uses this posterior as the prior for observation x or one obtains the posterior of $\theta|x, y$ directly.

2. Show that if z, a_1, a_2, b_1 and b_2 are scalars then

$$\frac{(z - a_1)^2}{b_1} + \frac{(z - a_2)^2}{b_2} = \frac{(z - c)^2}{d} + \frac{(a_1 - a_2)^2}{b_1 + b_2}$$

where $d^{-1} = b_1^{-1} + b_2^{-1}$ and $c = d(b_1^{-1}a_1 + b_2^{-1}a_2)$. Extend the result for the multivariate case with d-dimensional vectors z, a_1 and a_2 and $d \times d$ symmetric positive definite matrices B_1 and B_2 showing that

$$(z - a_1)'B_1^{-1}(z - a_1) + (z - a_2)'B_2^{-1}(z - a_2) =$$
$$(z - c)'D^{-1}(z - c) + (a_1 - a_2)'(B_1 + B_2)^{-1}(a_1 - a_2)$$

where $D^{-1} = B_1^{-1} + B_2^{-1}$ and $c = D(B_1^{-1}a_1 + B_2^{-1}a_2)$.

3. Consider a sample $x_1, ..., x_n$ from a distribution with density $f(x|\mu, \sigma)$ that admits a location-scale model (Exercise 1.5) with location parameter μ and scale parameter σ. Show that the non-informative prior is:

(a) $p(\mu) \propto k$ if σ is known;

(b) $p(\sigma) \propto 1/\sigma$ if θ is known;

(c) $p(\mu, \sigma) \propto 1/\sigma^2$ but reduces to $p(\mu, \sigma) \propto 1/\sigma$ if independent priors are chosen.

Hint: show that $I(\theta) = E\left[\left(\frac{\partial \log f(x|\theta)}{\partial \theta}\right)\left(\frac{\partial \log f(x|\theta)}{\partial \theta}\right)' |\theta\right]$.

4. Consider $x \sim bin(n, \theta)$. Obtain Jeffreys' prior for θ and show that despite being non-informative it is a proper distribution. Obtain the normalizing constant and draw the graph of its density.

5. Show that the multinomial distribution with index n and probabilities $p_1, ..., p_k$ satisfying $\Sigma_i x_i = n$ and $\Sigma_i p_i = 1$ belongs to the $k-1$-parameter exponential family with vector $\theta = (\theta_1, ..., \theta_{k-1})'$ of canonical parameters with components $\theta_i = \log(p_i/p_k)$, $i = 1, ..., k-1$. Obtain the functions $a(x)$ and $b(\theta)$.

6. Consider the situation of Example 2.1 but assume that the prior is a discrete mixture of two normal densities.

   (a) Show that this prior is still conjugate, that is, the posterior is a discrete mixture of two normal densities.

   (b) Generalize the result in (a) for discrete mixtures of k normal densities, $k = 2, 3, ...$

   (c) Generalize the result in (b) for discrete mixtures of k $CP(\alpha_i, \beta_i)$ distributions, $i = 1, ..., k$, and observations following a $EF(\mu)$ distribution.

   (d) Calculate the posterior for a prior given by the mixture $0.8N(0, 1) + 0.2N(5, 4)$ and the data used in Figure 2.1. Plot the posterior density and compare it with the densities in that figure.

7. Show that once the probability is fixed, HPD credibility intervals are always the shortest ones. Obtain (numerically if necessary) the HPD intervals for the posterior densities of θ considered in Figure 2.1 and for the posterior resulting from Exercise 2.6d.

8. Prove equation (2.13), that is, show that

$$\phi^{[(n_0+n+d)/2]-1} \exp\left\{-\frac{\phi}{2}[n_0 S_0 + S_e + (\beta - b_0)' B_0^{-1}(\beta - b_0) + Q(\beta)]\right\}$$

$$= \phi^{d/2} \exp\left\{-\frac{\phi}{2}(\beta - b_1)' B_1^{-1}(\beta - b_1)\right\} \phi^{(n_1/2)-1} \exp\left\{-\frac{\phi}{2}n_1 S_1\right\}$$

where $n_1 = n_0 + n$, $n_1 S_1 = n_0 S_0 + (y - X b_1)'y + (b_0 - b_1)' B_0^{-1} b_0$, $b_1 = B_1(B_0^{-1} b_0 + X'y)$ and $B_1^{-1} = B_0^{-1} + X'X$.

9. Consider the regression model $y|\beta, \phi \sim N(X\beta, \phi^{-1} I_n)$ and assume that the prior for (β, ϕ) has density proportional to

$$\phi^{\frac{n_0}{2}-1} \exp\left\{-\frac{1}{2}[(\beta - b_0)' B_0^{-1}(\beta - b_0) + \phi(\beta - b_1)' B_1^{-1}(\beta - b_1) + n_0 S_0 \phi]\right\}$$

   (a) Obtain the conditional prior distributions of $\beta|\phi$ and $\phi|\beta$.

   (b) Show that the above distribution is conditionally conjugate for β and for ϕ.

   (c) Assume now that the term $n_0 S_0 \phi$ in the expression of the prior density is replaced by $n_0 S_0/\phi$. Show that β is still conditionally conjugate but not ϕ.

10. Show that in the two-stage hierarchical model, the full conditional distribution of β_2 is $N(\mu_1, C_2^*)$ where $\mu_1 = C_2^*(C_2^{-1}\mu + X_2'\beta_1)$ and $C_2^* = (C_2^{-1} + X_2'X_2)^{-1}$.

11. Consider the K-stage hierarchical model

$$
\begin{aligned}
y \mid \beta_1, \phi &\sim N(X_1\beta_1, \phi^{-1}I_n) \\
\beta_k \mid \beta_{k+1}, \phi &\sim N(X_{k+1}\beta_{k+1}, \phi^{-1}C_k^{-1}), \quad k = 1, \cdots, K - 1 \\
\beta_K \mid \phi &\sim N(\mu, \phi^{-1}C_K^{-1}) \\
\phi &\sim G(n_0/2, n_0\sigma_0^2/2)
\end{aligned}
$$

   (a) Obtain the conditional prior distribution of $\beta_k \mid \phi$, $k = 1, ..., K - 1$.
   (b) Obtain the marginal prior distribution of β_k, $k = 1, ..., K - 1$.
   (c) Obtain the marginal distribution of y.
   (d) Obtain the conditional posterior distribution of $\beta_k \mid \phi$, $k = 1, ..., K-1$.
   (e) Obtain the marginal posterior distribution of β_k, $k = 1, ..., K - 1$.

12. Show that for the one-way analysis of variance model with hierarchical prior described in Example 2.7, $Cov(\beta_j, \beta_{j'} \mid \phi) = Cov(\beta_j, \beta_{j'}) = B_0$ and $Cor(\beta_j, \beta_{j'} \mid \phi) = Cor(\beta_j, \beta_{j'}) = (1 + \tau^2/B_0)^{-1}$.

13. (George, Makov and Smith, 1993) Consider observations $y_{ij} \sim EF(\mu_i)$, $j = 1, ..., n_i$, $i = 1, ..., d$, collected from d groups with different canonical parameters θ_i. The hierarchical prior for $\theta = (\theta_1, ..., \theta_d)'$ assume they form a sample from a $CP(\alpha, \beta)$ distribution and the second stage specifies a constant (possibly improper) prior for the hyperparameters α and β.

   (a) Obtain the full conditional posterior of θ and show it splits into a product of densities, each depending only on observations from one group and the hyperparameters. Apply the results to Poisson and Binomial observations.

   (b) Obtain the full conditional posterior of the hyperparameters and show that it is log-concave and proper. Apply the results to Poisson and Binomial observations.

14. Consider the model described in Example 2.8 with $\sigma_t^2 = \sigma^2$ and $W_t = W, \forall t$.

   (a) Show that the predictive distribution for an horizon $k > 0$ is

$$
y_{t+k}|D_t \sim N(m_t, C_t + kW + \sigma^2)
$$

   (b) Obtain the joint predictive distribution of $(y_{t+1}, \cdots, y_{t+k}|D_t)$.
   (c) Obtain the predictive distribution of the sum $y_{t+1}+...+y_{t+k}$ of future observations. In particular, show for $k = 2$ that

$$
y_{t+1} + y_{t+2}|D_t \sim N(2m_t, 4C_t + 2V + 5W).
$$

15. In the conditions of the previous exercise, show that the one-step-ahead filtered distribution for β is

$$\beta_{t-1}|D_t \sim N(m_{t-1}^t, C_{t-1}^t)$$

with

$$
\begin{aligned}
m_{t-1}^t &= m_{t-1} + (C_{t-1}/R_t)(m_t - m_{t-1}) \text{ and} \\
C_{t-1}^t &= C_{t-1} - (C_{t-1}/R_t)^2(R_t - C_t)
\end{aligned}
$$

16. In the conditions of Exercise 2.15, assume that observation y_t is missing and $D_t = D_{t-1}$. Obtain the distributions of $\beta_t|D_t$ and $y_{t+1}|D_t$ as functions of m_{t-1} and C_{t-1}.

17. Show that for dynamic models the full conditional distribution of β_t is $N(b_t, B_t)$ for $t = 1, ..., n$ where

$$
b_t = \begin{cases}
B_t(\sigma_t^2 F_t y_t + G_{t+1}'W_t^{-1}\beta_{t+1} + R^{-1}a) & , t = 1 \\
B_t(\sigma_t^2 F_t y_t + G_{t+1}'W_t^{-1}\beta_{t+1} + W_t^{-1}G_t\beta_{t-1}) & , t = 2, ..., n-1 \\
B_t(\sigma_t^2 F_t y_t + W_t^{-1}G_t\beta_{t-1}) & , t = n
\end{cases}
$$

and

$$
B_t = \begin{cases}
(\sigma_t^2 F_t F_t' + G_{t+1}'W_t^{-1}G_{t+1} + R^{-1})^{-1} & , t = 1 \\
(\sigma_t^2 F_t F_t' + G_{t+1}'W_t^{-1}G_{t+1} + W_t^{-1})^{-1} & , t = 2, ..., n-1 \\
(\sigma_t^2 F_t F_t' + W_t^{-1})^{-1} & , t = n
\end{cases}
$$

18. Show that for dynamic models

(a) $\pi(\beta_t|\beta_{t+1}, ..., \beta_n, y^n) = \pi(\beta_t|\beta_{t+1}, y^t)$ for $t = 1, ..., n-1$;

(b) $\pi(\beta_t, ..., \beta_n|y^n) = p(\beta_n|y^n) \times \prod_{k=t}^{n-1} p(\beta_k|\beta_{k+1}, y^t)$ for $t = 1, ..., n-1$;

(c) $\pi(\beta_t, \beta_{t+1}|y^n) = p(\beta_{t+1}|y^n)p(\beta_t|\beta_{t+1}, y^t)$ for $t = 1, ..., n-1$.

(d) $\beta_t|\beta_{t+1}, y^n \sim N(h_t^n, H_t^n)$ where

$$
\begin{aligned}
h_t^n &= m_t + C_t G_{t+1}'R_{t+1}^{-1}(\beta_{t+1} - a_{t+1}) \\
H_t^n &= C_t - C_t G_{t+1}'R_{t+1}^{-1}R_{t+1}R_{t+1}^{-1}G_{t+1}C_t
\end{aligned}
$$

(e) $\beta_t|y^n \sim N(m_t^n, C_t^n)$ where

$$
\begin{aligned}
m_t^n &= m_t + C_t G_{t+1}'R_{t+1}^{-1}(m_{t+1}^n - a_{t+1}) \\
C_t^n &= C_t - C_t G_{t+1}'R_{t+1}^{-1}(R_{t+1} - C_{t+1}^n)R_{t+1}^{-1}G_{t+1}C_t
\end{aligned}
$$

19. Show that for dynamic models the full conditional posterior distributions of the variances of the disturbances are $(\sigma^2|\beta, W) \sim IG(n_\sigma^*/2, n_\sigma^* S_\sigma^*/2)$ and $(W|\beta, \sigma^2) \sim IW(n_W^*/2, n_W^* S_W^*/2)$ with $n_\sigma^* = n_\sigma + n$, $n_\sigma^* S_\sigma^* = n_\sigma S_\sigma + \Sigma_t(y_t - F_t'\beta_t)^2$, $n_W^* = n_W + n - 1$ and $n_W^* S_W^* = n_W S_W + \Sigma_{t=2}^n(\beta_t - G_t\beta_{t-1})(\beta_t - G_t\beta_{t-1})'$.

## CHAPTER 3

Approximate methods of inference

3.1 Introduction

This chapter presents some of the methods proposed for Bayesian inference when the necessary calculations cannot be performed analytically. Some of these techniques are based on deterministic concepts while others are based on non-iterative simulation in opposition to the methods based on iterative simulation that form the core of this text. Therefore, only an introduction to the subject is presented. A more thorough treatment of the subject with comparisons and illustrations of the different techniques is given by Evans and Swartz (1995). The books by Carlin and Louis (1996), Gelman et al. (1995) and O'Hagan (1994) also provide nice reviews of the area with the first one also providing a summary of software available.

The main techniques presented in this chapter are normal and Laplace approximations based on asymptotics in section 3.2, quadrature approximations in section 3.3, Monte Carlo integration in section 3.4 and resampling techniques in section 3.5. The last two sections present solutions based on stochastic simulation. They generally involve sampling from an auxiliary distribution that serves different purposes in the context of each approximation.

The deterministic techniques rely upon approximate normality and asymptotic results in the sense of the sample size growing to infinity. These techniques were mostly developed during the last decade when the computational explosion that enabled computer-intensive methods to be performed was only starting. As will be seen, the complexity of the techniques increases substantially with the dimension of the parametric space. Similar comments are valid for the simulation techniques presented in this chapter. In particular, finding a suitable auxiliary distribution becomes an extremely difficult task. As a consequence, their application to a complete Bayesian analysis in complex models such as those presented at the last sections of the previous chapter is limited. Hierarchical and dynamic models have in common highly-dimensional parameter spaces that are difficult to approach for complete inference with the techniques presented in this chapter.

The last sections are more in the spirit of the book with the use of stochastic simulation for inference from the posterior distribution. The non-iterative form of the simulation used restricts its use in complex models with large numbers of parameters. For these cases, the use of iterative techniques based on Markov chains and described in the latter part of this book is almost inevitable.

Before going into the details of the techniques, it is important to recall that most summarization operations are provided by integration of the form

$$I = \int t(\theta)\pi(\theta)d\theta \tag{3.1}$$

The above expression provides the posterior mean of any transformation $\psi = t(\theta)$. When evaluating the posterior mean of θ, $t(\theta) = \theta$. When evaluating the posterior median c of a scalar θ, $t(\theta) = I(\theta < c)$, $I = 1/2$ and (3.1) is solved for c. Similarly, credibility regions are obtained by solving (3.1) for C with $t(\theta) = I(\theta \in C)$ and $I = 1 - \alpha$. The posterior variance matrix may be obtained by taking $t(\theta) = \theta\theta'$ and previously evaluating the posterior mean. Finally, for $\theta = (\theta_1, ..., \theta_d)'$ with components θ_i of any dimension, the marginal density of θ_i is given by (2.4). It can be rewritten as

$$\pi(\theta_i) = \int \pi(\theta_i|\theta_{-i})\pi(\theta_{-i})d\theta_{-i}$$

and again an integration over a posterior density is required with $t(\theta_{-i}) = \pi(\theta_i|\theta_{-i})$.

In very broad terms, experience gathered from previous authors suggests that deterministic techniques provide good results for low dimensional (say single digit) models. Beyond that, they become very complex to handle and Monte Carlo techniques have to be used. When the dimension of the model becomes increasingly large, then only Markov chain simulation seems to provide an adequate solution. Whenever possible analytical integration should be performed. This will reduce the dimension of the model where approximate methods are applied. Finally, it is important to mention that there is plenty of room for experimentation with combinations of these techniques.

3.2 Asymptotic approximations

The approximations here rely on results obtained when the sample size n gets large. Consider a parameter $\theta = (\theta_1, ..., \theta_d)'$ with posterior distribution π. Asymptotic approximations for π date back to the work of Laplace (1986) in the 18th. century. Nowadays, Laplace approximation is more commonly associated with the approximations presented in the final subsection. This nomenclature is kept in this book and the Laplace-motivated approximations of the next subsection are only referred to as normal approximations.

3.2.1 Normal approximations

These are based on a Taylor series expansion of the logarithm of the posterior density around the (assumedly unique) mode m

$$
\begin{aligned}
\log \pi(\theta) &= \log \pi(m) + \left[\frac{\partial \log \pi(m)}{\partial \theta}\right]' (\theta - m) \\
&\quad - \frac{1}{2!}(\theta - m)\left[-\frac{\partial^2 \log \pi(m)}{\partial \theta\, \partial \theta'}\right](\theta - m) + R(\theta) \\
&\doteq \log \pi(m) - \frac{1}{2}(\theta - m)\left[-\frac{\partial^2 \log \pi(m)}{\partial \theta\, \partial \theta'}\right](\theta - m) \quad (3.2)
\end{aligned}
$$

where the remainder $R(\theta)$ contains terms of order 3 or larger in the components of $(\theta - m)$ and is neglected in the approximation. Commonly, the posterior is known up to a proportionality constant, that is, only π^* is known where as before $\pi^*(\theta) = l(\theta)p(\theta) = k\pi(\theta)$ where $k = \int \pi^*(\theta)d\theta$. The expansion of $\log \pi^*$ around m gives analogously

$$
\begin{aligned}
\pi^*(\theta) &\doteq \pi^*(m) \exp\left\{-\frac{1}{2}(\theta - m)\left[-\frac{\partial^2 \log \pi^*(m)}{\partial \theta\, \partial \theta'}\right](\theta - m)\right\} \\
&= \pi^*(m) \exp\left\{-\frac{1}{2}(\theta - m)V^{-1}(\theta - m)\right\}
\end{aligned}
$$

and therefore

$$
k = \pi^*(m)(2\pi)^{d/2}|V|^{1/2} \tag{3.3}
$$

where

$$
V = \left[-\frac{\partial^2 \log \pi(m)}{\partial \theta\, \partial \theta'}\right]^{-1} = \left[-\frac{\partial^2 \log \pi^*(m)}{\partial \theta\, \partial \theta'}\right]^{-1}
$$

is minus the inverse of the curvature or Hessian or second derivative matrix of $\log \pi$ (and $\log \pi^*$) at the mode. Therefore, a posteriori, $\theta \sim N(m, V)$ and expectations evaluated with this approximation are denoted E_N. Approximate Bayesian inference proceeds as in the normal case with calculation of point estimates, quantiles, credibility intervals and marginal densities.

This approximation is similar to the asymptotic result of the maximum likelihood estimator, $\hat{\theta}$, if the prior is constant. In this case, $m = \hat{\theta}$ and V coincides with the inverse of the observed Fisher information matrix. As the sample size n increases, the likelihood becomes dominant and for n large, the influence of any non-degenerate prior density becomes negligible. In this case, a further approximation can be made by replacing the mode m by the maximum likelihood estimate $\hat{\theta}$ and V by the inverse of the observed Fisher information matrix evaluated at $\hat{\theta}$. Often, the expected (rather than the observed) Fisher information matrix is used. The normal approximation then becomes $\theta \sim N(\hat{\theta}, I^{-1}(\hat{\theta}))$ where $I(\theta)$ is given by (2.3). This approximation is sometimes called a Bayesian central limit theorem

(Carlin and Louis, 1996). All these approximations are $O(n^{-1})$. The above result is formally derived with a full discussion of the conditions for convergence to posterior normality by Heyde and Johnstone (1979).

Example 3.1 Consider the observation of $x \sim Poi(\lambda)$ and a conjugate prior $\lambda \sim G(\alpha, \beta)$. Example 2.3 shows that the posterior is $\pi(\lambda) = G(\alpha_1, \beta_1)$ where $\alpha_1 = \alpha + x$ and $\beta_1 = \beta + 1$. Hence, the posterior mean and variance are α_1/β_1 and α_1/β_1^2 respectively. The normal approximation to π has mean given by the posterior mode $m = (\alpha_1 - 1)/\beta_1$ and variance $V = (\alpha_1 - 1)/\beta_1^2$. The quality of the approximation can be assessed in Figure 3.1.

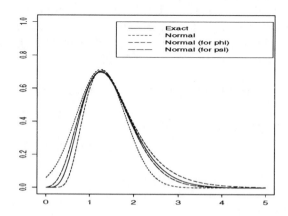

Figure 3.1. *Exact and approximated posterior densities for λ in Example 3.1 with* $\alpha = \beta = x = 3$.

Note that the normal approximation does not impose any condition on the parameter. It can be applied to any transformation $\psi = t(\theta)$. This will lead to a normal posterior with mean given by the posterior mode of ψ and variance given by minus the inverse of the Hessian of the log posterior of ψ evaluated at the mode. If the prior influence is discarded, the normal approximation for the posterior of ψ becomes $\psi \sim N(\hat{\psi}, I^{-1}(\hat{\psi}))$. Exercise 3.2 shows that this is equivalent to using the delta method for inference.

The normal approximation ignores skewness and secondary modes and it will only work well if the posterior is similar in shape to a normal distribution. When this does not apply, it is still possible to transform the parameter to a more adequate space. There are no optimal rules but good practical suggestions are reparametrizations that allow the parameter to vary over the real line and those leading to constant (Jeffreys) non-informative priors.

Example 3.1 (continued) A simple transformation taking λ to the real line is $\phi = \log \lambda$ whereas $\psi = \sqrt{\lambda}$ has constant non-informative prior. The respective normal approximations are $\phi \sim N(m_\phi, V_\phi)$ and $\psi \sim N(m_\psi, V_\psi)$ with $m_\phi = \log(\alpha_1/\beta_1)$, $V_\phi = \alpha_1^{-1}$, $m_\psi = \sqrt{(\alpha_1 - 1/2)/\beta_1}$ and $V_\psi = 1/4\beta_1$. These distributions imply different (non-normal) approximating distribution for λ. They improve over direct application of the normal approximation for λ and provide a reasonable approximation even for a small value of $n = 1$ as can be seen from Figure 3.1.

Approximations to $E(\psi)$ can be obtained as $E_N(\psi)$ or, when this is not possible, via the delta method. Further improvement is obtained by including second order terms in the Taylor series expansion of $\psi = t(\theta)$ around m so that

$$
\begin{aligned}
E[t(\theta)] \doteq E_N[t(\theta)] &\doteq E_N[t(m) + (\partial t(m)/\partial\theta)'(\theta - m) \\
&+ (\theta - m)'T(m)(\theta - m)/2] \\
&= t(m) + E_N\{tr[T(m)(\theta - m)(\theta - m)']/2\} \\
&= t(m) + tr\, T(m)V/2
\end{aligned}
$$

where T is the Hessian of $t(\theta)$. For scalar ψ and θ, the above approximation becomes $E[t(\theta)] = t(m) + t''(m)V/2$. These approximations are also $O(n^{-1})$.

Turning now briefly to multimodality, assume that, in addition to the global mode, a few local modes have been previously identified. Denoting the r modes by $m_1, ..., m_r$, the posterior density can be approximated by

$$
\pi(\theta) \doteq k \sum_{i=1}^{r} \pi^*(m_i) \exp\left\{-\frac{1}{2}(\theta - m_i)V_i^{-1}(\theta - m_i)\right\}
$$

where

$$
V_i = \left[-\frac{\partial^2 \log \pi^*(m_i)}{\partial\theta\,\partial\theta'}\right]^{-1} \text{ and } k^{-1} = (2\pi)^{d/2} \sum_{i=1}^{r} \pi^*(m_i)|V_i|^{1/2}
$$

This approximation provides $\theta \sim \Sigma_i k_i N(m_i, V_i)$ with $k_i = k\pi^*(m_i)|V_i|^{1/2}$, $i = 1, ..., r$.

3.2.2 Mode calculation

The above methods are simple and only involve the calculation of derivatives of the logarithm of the posterior density. The outstanding problem remains the calculation of the mode. It is generally obtained as the solution to the equation

$$
\frac{\partial \log \pi(\theta)}{\partial\theta} = 0
$$

In many cases, it is not possible to solve the equation analytically and numerical methods must be used. The most common ones are the Newton-

Raphson algorithm and its deterministic variants and Fisher scoring where the Hessian of $\log \pi$ is replaced by its sampling expectation. These are iterative methods that must be repeated until convergence is reached. A survey of the Newton-Raphson method in statistics is provided by Thisted (1988).

Example 3.2 Consider again generalized linear models with a normal prior $\beta \sim N(a, R)$ as in section 2.3.2. West (1985) showed that Fisher scoring leads at iteration j to the construction of *adjusted* observations $\tilde{y}_i = \tilde{y}_i(\beta^{(j-1)}) = g(\mu_i^{(j-1)}) + (y_i - \mu_i^{(j-1)})g'(\mu_i^{(j-1)})$ with variances $\tilde{V}_i = \tilde{V}_i(\beta^{(j-1)}) = -b''(\theta_i^{(j-1)})[g'(\mu_i^{(j-1)})]^2$, means $\mu_i^{(j-1)} = g^{-1}(x_i'\beta^{(j-1)})$ and canonical parameters $\theta_i^{(j-1)}$ satisfying $\mu_i^{(j-1)} = -b'(\theta_i^{(j-1)})$, $i = 1, ..., n$ where $\beta^{(j-1)}$ is the value of β at iteration $j - 1$. Forming the vector $\tilde{y} = (\tilde{y}_1, ..., \tilde{y}_n)'$ and the matrix $\tilde{V} = diag(\tilde{V}_1, ..., \tilde{V}_n)$, the *adjusted* regression model $\tilde{y}(\beta^{(j-1)})|\beta \sim N(X\beta, \tilde{V}(\beta^{(j-1)}))$ is obtained. Combining with the prior $\beta \sim N(a, R)$, the *adjusted* posterior is $\tilde{\pi}(\beta) = N(m^{(j)}, C^{(j)})$ where the posterior mean is $m^{(j)} = C^{(j)}(R^{-1}a + X'\tilde{V}^{-1}\tilde{y})$ and $C^{(j)} = (R^{-1} + X'\tilde{V}^{-1}X)^{-1}$. Then, one can set $\beta^{(j)} = m^{(j)}$ and move to iteration $j + 1$ by restarting the above procedure.

After convergence, $\beta^{(j)}$ is the posterior mode of β and it is also conceivable to approximate the posterior of β by a $N(m^{(j)}, C^{(j)})$. This method is an adaptation of the method of iterative reweighted least squares (IRLS) used to obtain the maximum likelihood estimator $\hat{\beta}$ in generalized linear models (McCullagh and Nelder, 1988). The main difference is the introduction of a prior distribution that leads to the calculation of the posterior mode instead of the maximum of the likelihood. The method provides as by-products an approximation for the variance and a normal approximation for the relevant distribution of β in the Bayesian framework and $\hat{\beta}$ in the frequentist framework. The extension of these algorithms to hierarchical and dynamic models is presented in Chapter 6.

In the case of more complex models with many components representing qualitatively different aspects of the model it is possible to use an iterative algorithm of mode search proposed by Lindley and Smith (1972) in the context of hierarchical models. Consider $\theta = (\phi, \psi)$ and let $\pi_\phi(\phi) = \max_\psi \pi(\phi, \psi) = \pi(\phi, \hat{\psi}(\phi))$ and $\pi_\psi(\psi) = \max_\phi \pi(\phi, \psi) = \pi(\hat{\phi}(\psi), \psi)$. If $\hat{\phi}$ $(\hat{\psi})$ maximizes π_ϕ (π_ψ) then it maximizes π and the value of θ that maximizes $\pi(\phi, \psi)$ is $\hat{\theta} = (\hat{\phi}, \hat{\psi})$. Note that $\hat{\phi}(\psi)$ $(\hat{\psi}(\phi))$ is the mode of conditional posterior of ϕ (ψ). Many times, it is easier to maximize the conditional densities than the joint density, especially under conditional conjugacy. Lindley and Smith (1972) suggest the following iterative algorithm:

1. initialize the iteration counter $j = 1$ and set an initial value $\phi^{(0)}$ for the mode of ϕ;

2. calculate the conditional modes $\psi^{(j)} = \hat{\psi}(\phi^{(j-1)})$ and $\phi^{(j)} = \hat{\phi}(\psi^{(j)})$;

3. change counter from j to $j+1$ and return to step 2 until convergence is reached.

The method moves towards the joint mode along the directions of the components. It may be slow if the posterior shows high correlation between the components. The convergence criterion can be any measure of distance between successive iterations. It is recommended to allow a few additional iterations to make sure that a point of maximum and not a saddlepoint has been reached. Calculation of posterior mode through iterations of conditional modes always converges to a mode. Additional verifications are required to ensure the mode encountered is global. It is recommended that a few different starting points are used to make sure that a global (and not local) maximum is reached. This method was also studied by Besag (1986) in the context of spatial statistics.

Example 3.3 (O'Hagan, 1994) Assume that $\begin{pmatrix} \phi \\ \psi \end{pmatrix} \sim N \left[\begin{pmatrix} 0 \\ 0 \end{pmatrix}, \begin{pmatrix} 1 & \rho \\ \rho & 1 \end{pmatrix} \right]$.
By (1.2), $\phi|\psi \sim N(\rho\psi, 1 - \rho^2)$ and $\psi|\phi \sim N(\rho\phi, 1 - \rho^2)$. The conditional modes are $\hat{\phi}(\psi) = \rho\psi$ and $\hat{\psi}(\phi) = \rho\phi$. Hence, $\phi^{(j)} = \rho\psi^{(j-1)} = \rho^2\phi^{(j-1)} = \rho^{2j}\phi^{(0)}$ and analogously $\psi^{(j)} = \rho^{2j}\psi^{(0)}$. This algorithm will converge to the unique joint mode $(0,0)$ for any value of ρ. The convergence can be slow if the correlation ρ between ϕ and ψ is close to 1 in absolute value.

The case of $\theta = (\theta_1, ..., \theta_d)'$ having d components becomes easier to understand. The iterative method calculates at each iteration the complete set of d full conditional modes $\hat{\theta}_1(\theta_{-1}), ..., \hat{\theta}_d(\theta_{-d})$. Starting with an initial value $\theta^{(0)}$, it iterates around full conditional modes until convergence is reached. It is illustrated below for the one-way model described in Example 2.7.

Example 3.4 Consider observations $y_{ij}|\beta_i, \sigma^2 \sim N(\beta_i, \sigma^2)$, $j = 1, ..., n_i$, and $\beta_i|\mu, \tau^2 \sim N(\mu, \tau^2)$, $i = 1, ..., d$. Assume independent priors for $\phi_1 = \sigma^{-2}$ and $\phi_2 = \tau^{-2}$ with respective distributions $G(n_\sigma/2, n_\sigma S_\sigma/2)$ and $G(n_\tau/2, n_\tau S_\tau/2)$. The joint distribution of all variables of the model $(y, \beta, \mu, \phi_1, \phi_2)$ is

$$\prod_{i=1}^{p} \prod_{j=1}^{n_i} f_N(y_{ij}; \beta_i, \phi_1^{-1}) \prod_{i=1}^{p} f_N(\beta_i; \mu, \phi_2^{-1}) f_N(\mu; b_0, B_0)$$

$$\times f_G(\phi_1; n_\sigma/2, n_\sigma S_\sigma/2) f_G(\phi_2; n_\tau/2, n_\tau S_\tau/2)$$

The following full conditional distributions are obtained:

a) $\pi(\beta_i|\beta_{-i}, \mu, \phi_1, \phi_2) = \pi(\beta_i|\mu, \phi_1) = N(\mu_i, C_i)$ where $\mu_i = C_i(n_i\phi_1\bar{y}_i + \phi_2\mu)$ and $C_i = (n_i\phi_1 + \phi_2)^{-1}$, $i = 1, ..., d$;

b) $\pi(\mu|\beta, \phi_1, \phi_2) = \pi(\mu|\beta, \phi_2) = N(b_1, B_1)$ where $b_1 = B_1(B_0^{-1}b_0 + \phi_2\bar\beta)$
and $B_1 = (B_0^{-1} + n\phi_2)^{-1}$ and $n = \Sigma_i n_i$.

c) $\pi(\phi_1|\beta, \mu, \phi_2) = \pi(\phi_1|\beta) = G(n_\sigma^*/2, n_\sigma^* S_\sigma^*/2)$ where $n_\sigma^* = n_\sigma + n$ and
$n_\sigma^* S_\sigma^* = n_\sigma S_\sigma + \Sigma_{i,j}(y_{ij} - \beta_i)^2$;

d) $\pi(\phi_2|\beta, \mu, \phi_1) = \pi(\phi_2|\beta, \mu) = G(n_\tau^*/2, n_\tau^* S_\tau^*/2)$ where $n_\tau^* = n_\tau + p$
and $n_\tau^* S_\tau^* = n_\tau S_\tau + \Sigma_i(\beta_i - \mu)^2$;

The respective modes of the distributions above are $\hat\beta_i = \mu_i$, $\hat\mu = b_1$,
$\hat\phi_1 = (n_\sigma^* - 2)/n_\sigma^* S_\sigma^*$ and $\hat\phi_2 = (n_\tau^* - 2)/n_\tau^* S_\tau^*$. The joint posterior mode
can be iteratively obtained through the following cycle:

1. initialize the iteration counter $j = 1$ and set initial values for the condi-
tional modes $\hat\beta_i^{(0)}$ and $\hat\mu^{(0)}$. For example, take $\beta_i^{(0)} = \bar y_i$ and $\mu^{(0)} = \bar{\bar y}$;

2a. evaluate the conditional modes $\phi_1^{(j)}$ and $\phi_2^{(j)}$ as functions of $\beta_i^{(j-1)}$ and
$\mu^{(j-1)}$;

2b. evaluate the conditional modes $\beta_i^{(j)}$ and $\mu^{(j)}$ as functions of $\phi_1^{(j)}$ and
$\phi_2^{(j)}$;

3. change the counter from j to $j+1$ and return to step 2 until convergence
is reached.

3.2.3 Standard Laplace approximation

One way to improve the approximations is to include higher order terms
in the Taylor series expansion of the log posterior. Lindley (1961, 1980)
considered the inclusion of third order terms leading to

$$\pi(\theta) \dot\propto \pi^*(m) \exp\left\{ -\frac{1}{2}(\theta - m)V^{-1}(\theta - m) + \frac{1}{3!}R(\theta) \right\}$$

where

$$R(\theta) = \sum_{i=1}^{p}\sum_{j=1}^{p}\sum_{k=1}^{p} \pi^{ijk}(\theta_i - m_i)(\theta_j - m_j)(\theta_k - m_k) \text{ and } \pi^{ijk} = \frac{\partial^3 \log \pi^*(m)}{\partial\theta_i\,\partial\theta_j\,\partial\theta_k}$$

The third order term has little influence for regions close to the mode but
away from the mode it becomes substantial and may cause instability in
the approximations. This can be alleviated by further replacing the term
$\exp\{R(\theta)/6\}$ by $1 + R(\theta)/6$ and the terms in the exponent will dominate the
tails. A new problem is caused by the possibility of negative approximation
values. Again, this is only likely to occur in the tails where the influence of
this term is negligible.

Integrals and posterior expectations can now be performed. The posterior
mean of a transformation $\psi = t(\theta)$ can be approximated by $E_N[t(\theta)(1 + R(\theta)/6)]$ and the term associated with $R(\theta)$ is clearly responsible for skewness corrections. For scalar θ, $R(\theta) = \pi^{(3)}(\theta - m)^3$ where $\pi^{(3)}$ is the third

derivative of $\log \pi$ evaluated at m. For $t(\theta) = \theta$, $E(\theta) \doteq m + \pi^{(3)} E_N(\theta - m)^4/6 = m + \pi^{(3)} V^2/2$.

Example 3.1 (continued) The posterior mean of λ can be approximated by the above expression. It is straightforward to obtain that $\pi^{(3)} = 2(\alpha_1 - 1)/m^3$ and that

$$
\begin{aligned}
\hat{E}(\lambda) &= m + \frac{2(\alpha_1 - 1)}{m^3} \frac{(\alpha_1/\beta_1^2)^2}{2} \\
&= \frac{\alpha_1 - 1}{\beta_1} + \frac{1}{\beta_1} \\
&= \frac{\alpha_1}{\beta_1}
\end{aligned}
$$

and the approximation is this case is exact.

For a vector θ,

$$
E[t(\theta)] = t(m) - \frac{1}{2} \sum_{i=1}^{d} \sum_{j=1}^{d} V_{ij} \left[\frac{\partial^2 t(m)}{\partial \theta_i \partial \theta_j} - \frac{\partial t(m)}{\partial \theta_i} \sum_{k=1}^{d} \sum_{l=1}^{d} V_{kl} \frac{\partial^3 \log \pi(m)}{\partial \theta_j \partial \theta_k \partial \theta_l} \right]
$$

The derivation is left as an exercise.

These approximations are called standard Laplace approximations by Kass, Tierney and Kadane (1988). They are $O(n^{-2})$ which is a qualitative improvement over the first order, normal-based approximations. Their main problem is the need to evaluate third order derivatives. It can be a substantial overhead if the dimension of θ is large. An alternative use of Laplace approximations that is also $O(n^{-2})$ but requires only calculations of first and second derivatives is presented below.

3.2.4 Exponential form Laplace approximations

Calculation of posterior means is given by (3.1) and can be rewritten as a ratio of integrals

$$
E[t(\theta)] = \int t(\theta)\pi(\theta)d\theta = \frac{\int t(\theta)l(\theta)p(\theta)d\theta}{\int l(\theta)p(\theta)d\theta} \tag{3.4}
$$

Equation (3.4) takes explicitly into account that the normalization constant may not be available, as is common in many posterior forms. Again the same idea of the previous section can be used to approximate integrands by quadratic forms in the log scale. The significant difference here is that it can be operated to both numerator and denominator in (3.4) which will be shown to lead to better approximations.

Assuming that $t(\theta) > 0$, define $L(\theta) = \log l(\theta) + \log p(\theta)$ and $L^*(\theta) = \log t(\theta) + \log l(\theta) + \log p(\theta)$, m^* as the value that maximizes L^* and V^* as minus the inverse Hessian of L^* at the point m^*. It follows that $E[t(\theta)] = \int \exp[L^*(\theta)]d\theta / \int \exp[L(\theta)]d\theta$.

Taking a Taylor series expansion of L^* up to the 2nd order term as in (3.2) for L gives

$$L^*(\theta) \quad = \quad L^*(m^*) - \frac{1}{2}(\theta - m^*)'V^{*-1}(\theta - m^*)$$

Replacing alongside (3.2) in (3.4) gives

$$E[t(\theta)] \doteq \left(\frac{|V^*|}{|V|}\right)^{1/2} \exp\left[L^*(m^*) - L(m)\right] \tag{3.5}$$

Tierney and Kadane (1986) noted that the advantage of this result is that despite it being based on approximations that are $O(n^{-1})$, the leading terms in the numerator and denominator cancel out. Therefore, the resulting approximation is $O(n^{-2})$.

Example 3.1 (continued) (Press, 1989) The posterior mean of λ may also be approximated by (3.5). In this case, $L(\lambda) = (\alpha_1 - 1)\log \lambda - \beta_1\lambda$ and $L^*(\lambda) = \alpha_1 \log \lambda - \beta_1\lambda$. Therefore, $m^* = \alpha_1/\beta_1$ and $V^* = \alpha_1/\beta_1^2$. Hence the approximation is

$$\hat{E}(\lambda) = \sqrt{\frac{\alpha_1/\beta_1^2}{(\alpha_1 - 1)/\beta_1^2}} exp\{(\alpha_1 \log m^* - \beta_1 m^*) - [(\alpha_1 - 1)\log m - \beta_1 m]\}$$

$$= \sqrt{\frac{\alpha_1}{\alpha_1 - 1}} \frac{(\alpha_1/\beta_1)^{\alpha_1} e^{-\alpha_1}}{[(\alpha_1 - 1)/\beta_1]^{\alpha_1 - 1} e^{-(\alpha_1 - 1)}}$$

$$= E(\lambda) \left(\frac{\alpha_1}{\alpha_1 - 1}\right)^{1/2} e^{-1}$$

When $\alpha = \beta = x = 3$, the percentual error is only 0.28%.

The calculations above are based on exponentiating the logarithm of the integrands. Kass, Tierney and Kadane (1988) refer to them as exponential form approximations to distinguish them from the standard form based on third derivatives. They improve upon the standard form by not requiring evaluations of third derivatives. The exponential form is usually preferred and for that reason is sometimes identified as *the* Laplace approximation.

The exponential form evaluates the logarithm of $t(\theta)$ and therefore can only be applied to positive functions $t(\theta)$. If t is not positive, it may be possible to define $t^*(\theta) = t(\theta) + a$ where a is a positive constant large enough to ensure that t^* is positive. Then, the exponential form can be used to approximate $E[t^*(\theta)]$. After that, it is easy to obtain $E[t(\theta)] = E[t^*(\theta)] - a$.

Alternatively, the moment generating function of $t(\theta)$ given by $M_t(s) = E[\exp(st(\theta))]$ can be used. The function M_t is always a positive function and can be approximated by $\hat{M}_t$ given by the exponential form (3.5) for any value of s. From basic probability theory, $E[t(\theta)] = dM_t(0)/ds$ and the posterior mean of $t(\theta)$ can be evaluated by differentiating $\hat{M}_t$ at 0. In practice, the analytic expression of $\hat{M}_t$ as a function of s is usually not available and numerical differentiation should be used. Tierney, Kass and Kadane (1989) showed that working with the moment generating function still retains approximation to $O(n^{-2})$ and that it equals the approximation based on addition of a large constant a when $a \to \infty$.

In the case of approximation of marginal densities, consider $\theta = (\phi, \psi)$ and assume the posterior marginal density of ϕ is required. It is given by $\pi(\phi) \propto \int \exp(L_\phi(\psi))d\psi$ where $L_\phi(\psi) = \log l(\phi, \psi) + \log p(\phi, \psi)$. Let $\hat{\psi}(\phi)$ be the value of ψ that maximizes L_ϕ. By the quadratic approximation (3.3),

$$\pi(\phi) \dot{\propto} \left| -\frac{\partial^2 L_\phi(\hat{\psi}(\phi))}{\partial \psi \partial \psi'} \right|^{-1/2} \exp[L(\phi, \hat{\psi}(\phi))]$$

If the prior $p(\phi, \psi) \propto k$ then $L = \log l$ and $\exp[L(\phi, \hat{\psi}(\phi))] = l(\phi, \hat{\psi}(\phi))$, the profile likelihood of ϕ. The resulting expression is the modified profile likelihood suggested by Cox and Reid (1987) for likelihood inference after elimination of the nuisance parameter ψ. Tierney and Kadane (1986) showed that the order of the approximation error falls to $O(n^{-3/2})$ which still compares favourably with errors of order $O(n^{-1/2})$ from the normal approximation.

A new problem here is the need for calculation of $\hat{\psi}(\phi)$ for a collection of values of ϕ. When ψ is conditionally conjugate, this value can usually be obtained analytically. The approximation for the marginal density may then be obtained in closed form. In many higher dimensional or non-conjugate models, this is rarely the case and maximizations must be repeated for each value of ϕ and the computational burden becomes substantial. When ϕ is uni- or bidimensional it is still possible to obtain graphical representations of its marginal density. For higher dimensions, there is also the additional problem associated with the cost of the calculation of the Hessian of L_ϕ.

The success of these approximations depends on unimodality of the posterior and similarity of the posterior with the normal forms used. Skew distributions are commonly associated with parameters with variation in a subset of the line. Again, useful reparametrizations may lead to a parameter with a more normal-like posterior. Hills and Smith (1992) suggest graphical techniques to guide the choice of the transformation required. Kass and Slate (1992) suggest analytical techniques based on third order derivatives of the posterior density to evaluate the similarity of the considered transformation with the normal distribution.

Example 3.5 Consider the model $x|\theta \sim bin(n, \theta)$ with Jeffreys prior $p(\theta) \propto \theta^{-1/2}(1 - \theta)^{-1/2}$. From Bayes' theorem

$$\pi(\theta) \propto \theta^x(1 - \theta)^{n-x}\theta^{-1/2}(1 - \theta)^{-1/2} = \theta^{x-1/2}(1 - \theta)^{n-x-1/2}$$

and, a posteriori, $E(\theta) = (x + 1/2)/(n + 1)$ can be exactly calculated. Assume the exact calculation was not possible and $E(\theta)$ is to be approximated by Laplace forms.

Maximization of L and L^* give $m = (x - 1/2)/(n - 1)$ and $m^* = (x + 1/2)/n$. Also, $V = n(x-1/2)(n-x-1/2)/(n-1)^3$ and $V^* = (x+1/2)(n-x-1/2)/n^2$. Replacing in (3.5) gives

$$\hat{E}(\theta) = \frac{(n-1)^{n+1/2}(x+1/2)^{x+1}}{n^{n+3/2}(x-1/2)^x}$$

When $n = 5$ and $x = 3$, $E(\theta) = 0.583$ and $\hat{E}(\theta) = 0.563$ with percentual error of 3.6%.

The reparametrization with constant Jeffreys prior is $\lambda = \arcsin(\sqrt{\theta})$ and $\theta = \sin^2\lambda$. The posterior becomes

$$\pi(\lambda) \propto (\sin^2\lambda)^x(1 - \sin^2\lambda)^{n-x}$$

Applying (3.5) for $t(\lambda) = \sin^2\lambda$ gives

$$\tilde{E}(\theta) = \frac{n^{n+1/2}(x+1)^{x+1}}{(n+1)^{n+3/2}x^x}$$

When $n = 5$ and $x = 3$, $\tilde{E}(\theta) = 0.580$ with percentual error of 0.6%.

These results were obtained by Achcar and Smith (1989) and provide an indication of the possible gains when calculations are performed in a suitable parametrization. Other parametrizations and values of n and x were considered as also leading to better results than those obtained with the original parametrization.

3.3 Approximations by Gaussian quadrature

Consider for the moment the one-dimensional problem of evaluation of $I = \int_a^b g(\theta)d\theta$. Quadrature rules approximate I by $\hat{I} = \sum_{i=1}^n w_ig(\theta_i)$ for some weights w_i and grid points θ_i, $i = 1, ..., n$. A simple rule is to take n equally spaced points with equal weights $c = (b - a)/n$. For most one-dimensional integrals, taking n of order 10^2 suffices for a good approximation. This rule can then be extended for integrals over the real line when $g(\theta)$ becomes negligible outside some wide limits. This is typically the case in statistical applications. Other simple quadrature rules are the trapezium rule where the grid endpoints θ_1 and θ_n receive half weight and Simpson's rule where weights alternate between $4c/3$ and $2c/3$ apart from the endpoints that receive weight $c/3$.

Improved rules were obtained by taking the form of the integrand into consideration. Gaussian rules were developed and tabulated when the integrand is well approximated by a form $h(\theta)p(\theta)$ where h is a polynomial function of θ and p is a density function. When $p = U[-1, 1]$, Gauss-Jacobi or Gauss-Legendre rules are obtained. When $p = G(\alpha, 1)$, Gauss-Laguerre rules are obtained and when $p = N(0, 1)$, Gauss-Hermite rules result. Each of the above rules is appropriate for a different support for θ. Given the important part played by the normal distribution in approximating the posterior density, the latter are usually preferred for approximating equations in the form (3.1). Reparametrizations must be made if the support of θ is not the real line.

Assuming an integration of the form $I = \int_{-\infty}^{\infty} \exp(-\theta^2) h(\theta) d\theta$, Gauss-Hermite rules approximate I by $\hat{I} = \Sigma_{i=1}^{n} w_i h(x_i)$ where x_i is the ith zero of the Hermite polynomial $H_n(\theta)$ and the respective weights are $w_i = 2^{n-1} n \sqrt{\pi} / [n H_{n-1}(x_i)]^2$, $i = 1, ..., n$. The approximation is exact and $\hat{I} = I$ if h is a polynomial function of degree $2n - 1$ or less. The approximation will be accurate if h is well approximated by a polynomial of degree $2n - 1$. Virtually any function can be well approximated for n sufficiently large. In practice, low values of n (around 6 to 8) are enough. The values of the weights w_i and zeros x_i can be found in mathematical handbooks such as Abramowitz and Stegun (1965).

For applications, adjustment must be made to account for a non-standard normal distribution. The integral $I = \int_{-\infty}^{\infty} g(\theta) d\theta$ where $g(\theta)$ is well approximated by $h(\theta) f_N(\theta; m, V)$ is given by $\hat{I} = \Sigma_i q_i h(\theta_i)$ where $q_i = w_i \sqrt{(2V)} e^{\theta_i^2}$ and $\theta_i = m + \sqrt{(2V)} x_i$, for weights w_i and zeros x_i, $i = 1, ..., n$ (Naylor and Smith, 1982). If the posterior $\pi(\theta) = k\pi^*(\theta)$ is known up to a proportionality constant k, Gauss rules can be applied to the numerator and denominator of (3.4) to approximate $E[t(\theta)]$ by

$$\frac{\sum_{i=1}^{n} q_i t(\theta_i) \pi^*(\theta_i)}{\sum_{i=1}^{n} q_i \pi^*(\theta_i)}$$

with m and V given by the posterior mean and variance of θ. Consequently, the rule also provides estimates of k. Note that the same approximation was applied to the numerator and denominator. If the posterior density is well approximated by a product of a normal by a polynomial of degree $2n - 3$ then evaluation of the constant k, the mean and the variance can be efficiently performed using a single n-point Gaussian rule.

In the multivariate case, consider integration of $g(\theta)$ where θ is a d-dimensional vector. Then

$$\int g(\theta) d\theta = \int \int g(\theta_1, \theta_{-1}) d\theta_{-1} d\theta_1$$

$$= \int g_1(\theta_1) d\theta_1$$

where $g_1(\theta_1) = \int g(\theta_1, \theta_{-1}) d\theta_{-1}$. The remaining integral of g_1 is one-dimensional and can be approximated by $\sum_{i_1=1}^{n_1} q_{i_1} g_1(\theta_{1,i_1})$ for weights q_{i_1} and points θ_{1,i_1} as before. Of course, g_1 itself is the result of a $(d-1)$-

dimensional integral that must be evaluated. For each of the n_1 points chosen for θ_1, it is given by

$$\int g(\theta_1, \theta_{-1}) d\theta_{-1} = \int \int g(\theta_1, \theta_2, \theta_{-12}) d\theta_{-12} d\theta_2$$

$$= \int g_2(\theta_1, \theta_2) d\theta_2$$

where $g_2(\theta_1, \theta_2) = \int g(\theta_1, \theta_2, \theta_{-1}) d\theta_{-12}$ and θ_{-12} is a $(d-2)$-dimensional vector obtained from θ by removal of its first and second components. Integration of g_2 for any given value of θ_1 is one-dimensional and can be approximated by $\sum_{i_2=1}^{n_2} q_{i_2} g_2(\theta_1, \theta_{2,i_2})$ for weights q_{i_2} and points θ_{2,i_2}. Therefore, I is approximated by

$$\hat{I} = \sum_{i_1=1}^{n_1} q_{i_1} g_1(\theta_{1,i_1})$$

$$= \sum_{i_1=1}^{n_1} q_{i_1} \sum_{i_2=1}^{n_2} q_{i_2} g_2(\theta_{1,i_1}, \theta_{2,i_2})$$

Proceeding by this route for the remaining $d-2$ one-dimensional integrals gives

$$\hat{I} = \sum_{i_1=1}^{n_1} q_{i_1} \cdots \sum_{i_d=1}^{n_d} q_{i_d} g(\theta_{1,i_1}, ..., \theta_{d,i_d})$$

The approximation involves evaluation of the function g at $\prod_i^d n_i$ (or n^d, when rules of same size n are used for all components) points.

This integration by Cartesian product rules is very sensitive to the choice of m and V and depends on an approximate assumption of independence between the components. If they are highly correlated, most of the evaluations of g will fall in regions of negligible value leading to very poor approximations (Smith et al., 1987). They suggest the use of an iterative strategy where the parameters are initially transformed to take values on the real line. Approximate posterior mean m and variance V for the resulting parameter vector ϕ are obtained. A new, standardized parameter ψ is constructed according to $\psi = L(\phi - m)$ where L is the square root of V^{-1} satisfying $LL' = V^{-1}$. ψ will then be a vector with approximately independent, standardized components. Cartesian product rules are then used to obtain new values of m and V. These are then used to redefine ψ and the iteration continues until the values of m and V converge. Naylor and Smith (1982) recommend that this iteration should be performed on a low dimension grid with n^d points and after convergence calculations should be refined with a larger grid with $(n+1)^d$ points. When convergence between different grid sizes is also achieved, posterior quantities of interest are then calculated. Iterations can start with n as low as 3.

An outstanding problem with these cartesian product rules is that the number of function evaluations increases exponentially with the dimension of θ. This limitation restricts the application of Cartesian product quadrature rules to models of single digit dimension. Alternative rules to avoid this so-called *curse of dimensionality* are discussed by Evans and Swartz (1988), Shaw (1988) and Dellaportas and Wright (1991).

3.4 Monte Carlo integration

Consider as before the problem of solving equation (3.1). Let $q(\theta)$ be another density for θ with the same support of π. Then

$$I = \int \frac{t(\theta)\pi(\theta)}{q(\theta)} q(\theta) d\theta = E_q \left[\frac{t(\theta)\pi(\theta)}{q(\theta)} \right]$$

where E_q denotes expectation with respect to q. If a sample $\theta_1, ..., \theta_n$ from q is available then

$$\hat{I} = \frac{1}{n} \sum_{i=1}^{n} \frac{t(\theta_i)\pi(\theta_i)}{q(\theta_i)} \tag{3.6}$$

is a method of moments estimator of I. It enjoys good frequentist properties such as:

- It is an unbiased estimator in that $E_q(\hat{I}) = I$.
- Its variance is in the form $V_q(\hat{I}) = \sigma^2/n$ where $\sigma^2 = \int [t^2(\theta)\pi^2(\theta)/q(\theta)]d\theta - I^2$.
- It has a central limit theorem stating that

$$\sqrt{n}\frac{\hat{I} - I}{\sigma} \to N(0, 1) \text{ as } n \to \infty \tag{3.7}$$

in distribution.

- It is a strongly consistent estimator of I in that

$$\hat{I} \to I \text{ as } n \to \infty \tag{3.8}$$

with probability 1.

The classical nature of the above results leads to objections by Bayesians (O'Hagan, 1987). These results provide important messages, however, and in practice they are widely used. Strong consistency follows directly from the strong law of large numbers (Feller, 1968). So increasing the size n of the sample from q will lead to a virtually error-free estimation at rate $O(n^{-1/2})$. Unlike asymptotic results, this value of n is under the control of the researcher and can be increased by drawing more values from q. The constant σ^2 depends on q and can also be estimated by the methods of moments.

The generating density q is usually called the importance density and sampling from q is called importance sampling. There are no restrictions on

q and the simplest choice is the uniform distribution when the support of θ is compact. It can be shown that the optimal choice in terms of minimizing σ^2 and hence the estimation error is to take $q \propto t \times \pi$. Unfortunately, for most cases where (3.1) cannot be evaluated analytically, it will be very difficult to draw samples from π. The above results however suggest that q should be taken as close as possible to π but still available for easy sampling. In any case, the importance density q can be chosen to approximate $t \times \pi$ for each required expectation of $t(\theta)$ or can be chosen to be the same for all integrations of interest. Kloek and van Dijk (1978) recommend the latter with the importance density q chosen to approximate π.

Geweke (1989) provides a formal proof of the central limit theorem. It may be used to assess coverage probabilities by confidence intervals thus providing error bounds for the estimates unlike the previous estimates proposed. Carlin and Louis (1996) and Evans and Swartz (1995) consider this ability as one of the main strengths of approximations based on Monte Carlo techniques.

A problem that usually arises in Bayesian applications is that π is only known up to a proportionality constant. Posterior expectations are really a problem involving a ratio of two integrals as pointed out in (3.4). The resulting approximation is based on the ratio of two Monte Carlo estimators of integrals. Using the same importance density q as recommended above, the numerator and the denominator are approximated by (3.6) with π replaced by $\pi^* = l \times p$ and in the case of the denominator $t = 1$. The form of the estimator is then

$$\hat{I} = \frac{\sum_{i=1}^{n} t(\theta_i)\pi^*(\theta_i)/q(\theta_i)}{\sum_{i=1}^{n} \pi^*(\theta_i)/q(\theta_i)}$$

where the θ_is are the same on numerator and denominator and are sampled from q. The above estimator is only asymptotically unbiased but is still a strongly consistent estimator of I.

Monte Carlo integration has been connected to Bayesian inference after its introduction in applied Econometrics by Kloek and van Dijk (1978). Medium sized models have been commonly used in this area and their paper showed it is a viable technique. Much effort has been devoted since then to the specification of suitable importance density functions. It is important that it matches π as close as possible and that it blankets π in the tails. Otherwise, the very few points sampled in the tails may have large contributions to $\hat{I}$ and estimates will be unstable. This suggests that normal distributions should be avoided if possible.

In the multivariate setting, natural choices for importance density are the Student's t distributions with low degrees of freedom. These are easy to sample, have thick tails and support over R^d. They may therefore require transformation of some of the components of θ to the real line. Geweke (1989) suggest the use of split-t distributions which are obtained by rescaling each component of θ differently for positive and negative values to accommodate skewness. Oh and Berger (1993) suggest the use of mixtures of t-distributions to accommodate posterior multimodality.

These functions require specification of mean and variance which themselves are obtained by integration. This suggest an iterative scheme where means and variances are evaluated for a given importance function and used to update mean and variance specifications of a new importance function. The process is repeated until the successive values of means and variances do not change. Then, integrations of interest are performed. Adaptive strategies have been suggested by Kloek and van Dijk (1978) and Smith et al. (1987). Oh and Berger (1992) establish convergence results of these iterative strategies. Examples in Evans and Swartz (1995) suggest that few (less than 10) iterations are required for convergence with sample sizes of order 10^3 to 10^4.

3.5 Methods based on stochastic simulation

Methods of inference based on stochastic simulation use samples from the posterior π to summarize information. Obviously, Monte Carlo integration is also a method of inference based on simulation. The difference lies in the fact that to perform the necessary integrations there was no need to actually produce a sample from π. This section explores methods designed to (at least approximately) produce a sample from π and shows the use of these methods in Bayesian inference. This is the theme of this book and this section provides the first formal contact with it.

From the start it should be pointed out that no matter how large the sample is it is only a partial substitute for the information contained in the expression of a density. These methods provide only an approximation to the posterior and should only be used when it is not possible to extract information from the posterior analytically. As approximations, they share the advantages of Monte Carlo integration techniques: their accuracy is controlled by the size of the sample irrespective of the number of observations in the data and the approximation errors can be probabilistically measured.

Once a sample $\theta_1, ..., \theta_n$ from π is available, many standard summarization operations can be approximately made. The posterior mean of θ is estimated by the arithmetic average of sample values. Posterior moments of θ and more generally posterior means of transformations $t(\theta)$ are estimated by the average of the corresponding operations on the θ_i. For example, if $\theta = (\theta_1, ..., \theta_d)'$, the posterior mean of the parameter $\psi = \theta_1^2 + ... + \theta_d^2$ is estimated by $\hat{\psi} = (1/n) \sum_{i=1}^{n} \psi_i$ where $\psi_i = \theta_{i1}^2 + ... + \theta_{id}^2$ and $\theta_i = (\theta_{i1}, ..., \theta_d)'$ is the ith element of the sample, $i = 1, ..., n$. Posterior probabilities of intervals or regions C are simply approximated by the proportions of sample values belonging to C. A $100(1 - \alpha)$ % credibility interval for a scalar transformation ψ of θ is approximately given by $[\psi_{([\alpha n/2])}, \psi_{([n(1-\alpha/2)])}]$ where $x_{(j)}$ is the jth ordered sample value and $[x]$ is the smallest integer not larger than x. For example, if $n = 1000$ and $\alpha = 0.05$, the credibility interval will have endpoints given by the 25th and 975th largest values of ψ. Of course, unequal tail probabilities can also be used.

All the above approximations are unbiased, method of moments estimates of their respective posterior quantities and their sampling error can be assessed (Exercise 3.14). Also, once a sample from a scalar quantity is available, a histogram can be plotted and provides an estimate of the density. These histograms provide a rough, discrete-like estimate that is often adequate to convey relevant aspects of the posterior. However, they may be smoothed according to some non-parametric smoothing density estimation technique (Silverman, 1986). A Bayesian approach to the problem is given by West (1992).

This approach can be readily applied to obtain samples from a predictive distribution. Recall from Chapter 2 that the predictive distribution for y after observing x is given by $f(y|x)$ which itself is the marginal density obtained from the joint posterior $f(y, \theta|x) = f(y|\theta)\pi(\theta)$. A sample $y_1, ..., y_n$ from this distribution can be obtained via mixture techniques (section 1.3) by drawing each y_i from $f(y|\theta_i)$ where θ_i is the ith element of the sample from the posterior π, $i = 1, ..., n$.

Availability of a sample for $\theta = (\theta_1, ..., \theta_d)'$ implies that samples for the component θ_j, $j = 1, ..., d$, are given by the jth components of each sample value θ_i, $i = 1, ..., n$. This trivial operation replaces the often intractable integration of the posterior over the remaining $d-1$ components of θ. More generally, a sample of any parametric transformation $\psi = t(\theta)$ is given by $\psi_1, ..., \psi_n$ where $\psi_i = t(\theta_i)$, $i = 1, ..., n$.

The previous sections showed the importance reparametrizations can have in techniques for approximate inference. It is nice to know that these can be easily handled by simulation-based methods. In some circumstances, although the problem is parametrized by θ, it is more convenient to proceed sampling through $\psi = t(\theta)$, where t is a one-to-one transformation. Once a sample $\psi_1, ..., \psi_n$ is obtained, a sample from θ is given by $\theta_1, ..., \theta_n$ where $\theta_i = s(\psi_i)$ and s is the inverse of the function t.

This easy operation replaces possibly difficult operations required to obtain the density of transformations. These densities can only be obtained when an explicit expression for the transformation t is available. Sampling-based methods can do without it. Press (1989) considers sampling from the distribution of the second largest eigenvalue λ_2 of a random matrix Ψ. There is no explicit analytic expression form for t such that $\lambda_2 = t(\Psi)$. Nevertheless, such a function exists and is well defined. Furthermore, it can be used to obtain λ_2 for any given value of Ψ by numerical techniques. So, if a sample $\Psi_1, ..., \Psi_n$ is available, so will be a sample $\lambda_{21}, ..., \lambda_{2n}$. Sampling-based approximate inference will then become possible when analytic methods are hopeless.

All the above derivations are based on a sample from π. Unfortunately, it is very complicated to directly sample from π for the vast majority of problems of practical relevance. Therefore, some ingenuity must be exercised to allow such a task to be accomplished. This book is concerned with stochastic simulation via Markov chains but it is important that less elaborate, more limited techniques are described. These are proving to be useful in many problems where the parameter is not highly dimensional. This section provides a brief introduction to them.

Methods of generation from a density of interest π using samples from an auxiliary density q were presented in section 1.5. Smith and Gelfand (1992) propose simple forms to give this problem a Bayesian interpretation. Their idea was to provide a simulation-based parallel to Bayes' theorem. This can be achieved by taking q as the prior for θ.

Assume that the posterior density π is only known up to a proportionality constant. Therefore, $\pi^*(\theta) = l(\theta)p(\theta)$ is available where l is the likelihood and p is the prior for θ but $\pi = k\pi^*$ is not because the value of k is unknown. The methods of that section worked just as well with π^*.

3.5.1 Bayes' theorem via the rejection method

A value is drawn from π by the rejection method by generating a value from q and accepting it with probability $\pi^*(\theta)/Aq(\theta)$ (section 1.5.1). The constant $A < \infty$ satisfies the blanketing condition $\pi^*(\theta)/q(\theta) < A$, for all θ. Efficiency of the method is improved as the value of A gets smaller.

Taking $q = p$, the prior for θ, and $\pi^* = l \times q$ gives $\pi^*(\theta)/q(\theta) = l(\theta)$ and the smallest constant A ensuring an envelope is $l_{max} = \max_\theta l(\theta) = l(\hat{\theta})$ where $\hat{\theta}$ is the maximum likelihood estimator of θ. The acceptance ratio becomes

$$\frac{\pi^*(\theta)}{Aq(\theta)} = \frac{l(\theta)p(\theta)}{l_{max}p(\theta)} = \frac{l(\theta)}{l_{max}}$$

which does not exceed 1.

Bayes' theorem describes the passage from prior to posterior after observing the data. A sampling version of the theorem describes the passage of a sample from the prior to a sample from the posterior. By the rejection method, this passage is clear:

1. A value θ is drawn from the prior density $p(\theta)$;

2. This value is accepted with probability $l(\theta)/l_{max}$.

Some care must be exercised despite the simplicity of the method. First, the method always ends up with a resample smaller than or equal to the original sample as some values proposed in Step 1 will not be accepted in Step 2. This can be a problem when sequentially updating the prior as in dynamic models or when prior and likelihood provide conflicting information. The prior density and likelihood functions that led to one of the posterior densities of Figure 2.1 are reproduced in Figure 3.2. Values generated according to the prior will have very reduced likelihood values and therefore will have low acceptance probabilities. This problem is more likely to occur in higher dimensional parameter spaces. Some study of the possible disagreement between prior and likelihood must be undertaken before a blind use of the method. Tachibana (1995) discusses forms to avoid or alleviate these problems (see also Newton and Raftery, 1994).

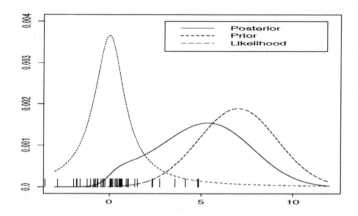

Figure 3.2. *Prior, likelihood and posterior corresponding to the posterior of Figure 2.1 associated with the Cauchy prior. The small bars still indicate a sample drawn from the prior.*

Another problem associated with the method is the need to maximize the likelihood to ensure total envelope. This maximization is not always easy to perform (section 3.2) and sometimes requires the use of sophisticated tools. This may render computationally expensive a method whose main attraction is simplicity. Below, an alternative method that does not require likelihood maximization is presented.

3.5.2 Bayes' theorem via weighted resampling

A sample from π can be generated via weighted resampling methods (section 1.5.2) by drawing a sample $\theta_1, ..., \theta_n$ from q and resampling from the discrete distribution in $\{\theta_1, ..., \theta_n\}$ with probabilities w_i given by

$$w_i = \frac{\pi(\theta_i)/q(\theta_i)}{\sum_{j=1}^{n} \pi(\theta_j)/q(\theta_j)} \quad , \quad i = 1, ..., n$$

Substituting the values of $q = p$ and noting that $\pi/q = kl$ gives $w_i = l(\theta_i)/\sum_{j=1}^{n} l(\theta_j)$, $i = 1, ..., n$. Therefore, Bayes' theorem via weighted resampling methods proceeds as follows:

1. A sample $\theta_1, ..., \theta_n$ is drawn from the prior density $p(\theta)$;

2. The values of the sample are resampled with probabilities

$$w_i = \frac{l(\theta_i)}{\sum_{j=1}^{n} l(\theta_j)} \quad , \quad i = 1, ..., n$$

Compared with rejection methods, this method does not necessarily reduce the original sample size and does not require maximizations. This only makes more relevant the need for care in case of conflicting information between prior and likelihood. In this case, many original sample values are equally unlikely under the posterior but this is not considered. Calculation of the weights only considers them relative to other sample points without accounting for their absolute posterior weight. Once again, higher dimensional parameter spaces will make difficult an efficient use of the method as adequate coverage of the relevant posterior regions by the original sample becomes harder to achieve.

These procedures can also be subject to an adaptive iterative scheme where original samples are initially formed with the prior. If this gives indications of inadequacy, the original sampling density is replaced by another one that places more weight in the relevant posterior regions. Smith and Gelfand (1992) also consider this idea. More discussion on the subject with applications to practical problems can be found in Stephens and Smith (1992), Mendes (1995), Tachibana (1995) and Moreira, Lopes and Schmidt (1996).

3.5.3 Application to dynamic models

This section is concluded with an application of sampling-based approaches to inference in dynamic models (section 2.5). It was possible to obtain the evolution, prediction and updating equations analytically under linearity and normality of the disturbances. Consider now general dynamic models with observation and system equations given by

$$
\begin{aligned}
y_t &= F_t(\beta_t, \epsilon_t) \text{ where } \epsilon_t \sim F_\epsilon & (3.9) \\
\beta_t &= G_t(\beta_{t-1}, w_t) \text{ where } w_t \sim F_w & (3.10)
\end{aligned}
$$

where F_ϵ and F_w are any (possibly non-normal) distributions. These distributions may in fact depend on t and the system equation may also depend on previous βs without invalidating the calculations below. Nevertheless, equations (3.9)-(3.10) are retained for notational simplicity.

Sequential inference for a dynamic model is based on obtaining for each time t the prior, predictive and updated distribution for the system parameter β_t. The first two distributions are

$$p(\beta_t|y^{t-1}) \;=\; \int p(\beta_t|\beta_{t-1})p(\beta_{t-1}|y^{t-1})d\beta_{t-1} \qquad (3.11)$$

$$p(y_t|y^{t-1}) \;=\; \int p(y_t|\beta_t)p(\beta_t|y^{t-1})d\beta_t \qquad (3.12)$$

and the last one is obtained via Bayes' theorem as

$$p(\beta_t|y^t) \propto p(y_t|\beta_t)p(\beta_t|y^{t-1}) \qquad (3.13)$$

None of these operations can be completely performed analytically for the generality assumed for the model and an alternative is to use a sampling-based approach to perform them approximately. A sequential simulation procedure follows the steps below:

1. Assume $\beta_0^{(1)}, ..., \beta_0^{(m_0)}$ is a sample of β_0 and set the time counter $t = 1$.

2. Evolution: obtain a sample $\beta_t^{*(1)}, ..., \beta_t^{*(m_{t-1})}$ from $p(\beta_t|y^{t-1})$ with $\beta_t^{*(j)} = G_t(\beta_{t-1}^{(j)}, w_t^{(j)})$ by drawing $w_t^{(j)}$ from F_w independently, $j = 1, ..., m_{t-1}$.

3. Prediction: obtain a sample $y_t^{(1)}, ..., y_t^{(m_{t-1})}$ from $p(y_t|y^{t-1})$ with $y_t^{(j)} = F_t(\beta_t^{*(j)}, \epsilon_t^{(j)})$ by drawing $\epsilon_t^{(j)}$ from F_ϵ independently, $j = 1, ..., m_{t-1}$.

4. Updating: obtain a sample $\beta_t^{(1)}, ..., \beta_t^{(m_t)}$ from $p(\beta_t|y^t)$ using any of the sampling-based versions of Bayes' theorem to (3.13).

5. Increase t to $t + 1$ and return to 2 until all time points have been processed.

The generations in 2 and 3 follow respectively from (3.11) and (3.12) using results about sampling from mixtures (section 1.3). Generations in 4 can be made either via rejection (Muller, 1991a) or weighted resampling (Gordon, Salmond and Smith, 1993). For both methods there are problems associated with the reduction of the sample through time and treatment of conflict between prior and likelihood. This conflict occurs with outlying observations that can indicate either an occasional or structural change in the pattern of the series. The above papers discuss possible treatments of these problems. The methods are somewhat restricted by not evaluating smoothing distributions and requiring draws from F_ϵ and F_w. In general, these distributions depend on unknown hyperparameters such as variance terms. Sampling-based approaches taking all unknown quantities into consideration will be presented in Chapters 5 and 6.

3.6 Exercises

1. Consider a posterior for λ in the conditions of Example 3.1 given by a $G(\alpha_1, \beta_1)$ distribution. Define the transformations $\phi = \log \lambda$ and $\psi = \sqrt{\lambda}$.

(a) Show by direct likelihood calculation and by exploring transformations on Fisher information that the Jeffreys non-informative prior for ψ is constant.

(b) Obtain the posterior distributions of ϕ and ψ and plot them for the values used in Figure 3.1.

(c) Obtain the normal approximations to the posterior distributions of ϕ and ψ.

(d) Obtain the expressions for the non-normal posterior densities of λ derived from the normal approximations to the posterior distributions of ϕ and ψ.

(e) Obtain the Lindley approximation for the posterior mean of ϕ and ψ.

2. The delta method is a well known approximation technique based on first order Taylor series expansions that finds applications in both frequentist and Bayesian inference. It states that if a random vector x has vector mean a and variance matrix B and $t(x)$ is a differentiable function then $t(x)$ has approximate vector mean $t(a)$ and variance matrix $(\partial t(a)/\partial x)'B(\partial t(a)/\partial x)$. If, in addition, x is normal then $t(x)$ will also be approximately normal.

(a) Use Taylor series expansions up to the linear term to prove the delta method.

(b) Show that application of the delta method to a transformation $\psi = t(\theta)$ with the normal approximation for the posterior of θ discarding the prior influence coincides with the direct application of the normal approximation for the posterior of ψ discarding the prior influence.

3. (Lindley, 1980) Show that the standard Laplace approximation for the posterior mean of $\psi = t(\theta)$ is given by

$$E(\psi) \doteq t(m) - \frac{1}{2}\sum_{i=1}^{d}\sum_{j=1}^{d}V_{ij}\left[\frac{\partial^2 t(m)}{\partial\theta_i\partial\theta_j} - \frac{\partial t(m)}{\partial\theta_i}\sum_{k=1}^{d}\sum_{l=1}^{d}V_{kl}\frac{\partial^3\log\pi(m)}{\partial\theta_j\partial\theta_k\partial\theta_l}\right]$$

Show also that if θ_l is the lth component of θ then

$$E(\theta_l) = m_l + \frac{1}{2}\sum_{i=1}^{d}\sum_{j=1}^{d}\sum_{k=1}^{d}\frac{\partial^3\log\pi(m)}{\partial\theta_j\partial\theta_k\partial\theta_l}V_{ij}V_{kl}$$

4. Go through the Fisher scoring algorithm to evaluate the maximum likelihood estimator of β in a generalized linear model as described for example in McCullagh and Nelder (1988) and show that it coincides with the posterior mode when a constant prior is used for β by altering the relevant steps in Example 3.2.

5. Show that in the one-way classification model with a two-stage hierarchical prior described in Example 3.4, the full conditional distributions are:

   (a) $\pi(\beta_i|\beta_{-i}, \mu, \phi_1, \phi_2) = \pi(\beta_i|\mu, \phi_1) = N(\mu_i, C_i)$ where $\mu_i = C_i(n_i\phi_1\bar{y}_i + \phi_2\mu)$ and $C_i = (n_i\phi_1 + \phi_2)^{-1}$, $i = 1, ..., p$;

   (b) $\pi(\mu|\beta, \phi_1, \phi_2) = \pi(\mu|\beta, \phi_2) = N(b_1, B_1)$ where $b_1 = B_1(B_0^{-1}b_0 + \phi_2\bar{\beta})$, $B_1 = (B_0^{-1} + n\phi_2)^{-1}$ and $n = \Sigma_i n_i$.

   (c) $\pi(\phi_1|\beta, \mu, \phi_2) = \pi(\phi_1|\beta) = G(n_\sigma^*/2, n_\sigma^*S_\sigma^*/2)$ where $n_\sigma^* = n_\sigma + n$ and $n_\sigma^*S_\sigma^* = n_\sigma S_\sigma + \Sigma_{i,j}(y_{ij} - \beta_i)^2$;

   (d) $\pi(\phi_2|\beta, \mu, \phi_1) = \pi(\phi_2|\beta, \mu) = G(n_\tau^*/2, n_\tau^*S_\tau^*/2)$ where $n_\tau^* = n_\sigma + p$ and $n_\tau^*S_\tau^* = n_\tau S_\tau + \Sigma_i(\beta_i - \mu)^2$.

6. Let $\theta = (\phi, \psi)$ where ϕ and ψ are d-dimensional and r-dimensional parameter vectors and θ has density π.

   (a) Obtain the normalization constants of the Laplace approximation for the marginal densities of ϕ and ψ.

   (b) Let $\pi(\theta) = NG_d(\mu, \sigma^2, n, S)$ and, therefore, $r = 1$. Show that the Laplace approximations for the marginal densities of ϕ and ψ coincide with the exact Student's t and Gamma densities.

7. (Achcar and Smith, 1989) Consider Example 3.5 from section 3.3.

   (a) Deduce the expressions of $\hat{E}(\theta)$ and $\tilde{E}(\theta)$.

   (b) Using the logistic transformation $\phi = \text{logit}(\theta)$, obtain another approximation for $E(\theta)$ using the fully exponential form of Laplace approximation and compare it with the approximations obtained for $n = 5$ and $x = 3$.

8. Consider two independent binomial observations $x_i|\theta_i \sim bin(n_i, \theta_i)$, $i = 1, 2$, with non-informative prior $p(\theta_1, \theta_2) \propto \theta_1^{-1/2}(1 - \theta_1)^{-1/2}\theta_2^{-1/2}(1 - \theta_2)^{-1/2}$ and define $\psi = \theta_1/\theta_2$.

   (a) Show that the posterior mean of ψ is $E(\psi) = n_2(x_1 + 1/2)/[(n_1 + 1)(x_2 - 1/2)]$;

   (b) Using the exponential form Laplace approximation, show that

$$E(\psi) \doteq \frac{(n_1 - 1)^{n_1+1/2}(n_2 - 1)^{n_2+1/2}(y_1 + 1/2)^{y_1+1}(y_2 - 3/2)^{y_2-1}}{(n_1)^{n_1+3/2}(n_2 - 2)^{n_2-1/2}(y_1 - 1/2)^{y_1}(y_2 - 1/2)^{y_2}}$$

   (c) Using the parametrization $\lambda_i = \text{arc} \sin(\sqrt{\theta_i})$, $i = 1, 2$, obtain the approximation

$$E(\psi) \doteq \frac{(2n_1)^{(2n_1+1)/2}(2n_2)^{(2n_2+1)/2}(2y_1 + 2)^{y_1+1}(2y_2 - 2)^{y_2-1}}{(2n_1 + 2)^{(2n_1+3)/2}(2n_2 - 2)^{(2n_2-1)/2}(2y_1)^{y_1}(2y_2)^{y_2}}$$

(d) Using now the parametrization $\phi_i = \text{logit}(\theta_i)$, $i = 1, 2$, obtain the approximation

$$E(\psi) \doteq \frac{(n_1 + 1)^{n_1+1/2}(n_2 + 1)^{n_2+1/2}(y_1 + 3/2)^{y_1+1}(y_2 - 1/2)^{y_2-1}}{(n_1 + 2)^{n_1+3/2}(n_2)^{n_2-1/2}(y_1 + 1/2)^{y_1}(y_2 + 1/2)^{y_2}}$$

(e) Consider a small set of values for x_1, x_2, n_1 and n_2 and evaluate the percentual errors of the approximations above.

9. Consider the integral $I = \int_{-\infty}^{\infty} g(\theta)d\theta$ where $g(\theta)$ is well approximated by $h(\theta)f_N(\theta; m, V)$. Show that the adjustment to the Gauss-Hermite rules to account for a non-standard normal distribution gives the approximation $\hat{I} = \Sigma_i q_i h(\theta_i)$ where $q_i = w_i \sqrt{(2V)}e^{-\theta_i^2}$ and $\theta_i = m + \sqrt{(2V)}x_i$, for weights w_i and zeros x_i, $i = 1, ..., n$.

10. Show that if θ is scalar, application of different 1-point Gauss-Hermite quadrature rules ($w_1 = \sqrt{(2\pi)}$ and $x_1 = 0$) to the numerator and denominator in (3.4) gives the exponential form Laplace approximation (3.5).

11. Generate a sample $x = (x_1, ..., x_n)$ of size $n = 50$ from the $N(\mu, \sigma^2)$ distribution where $\mu \sim N(\mu_0, \tau^2)$ and $\sigma^2 \sim IG(n_0/2, S_0/2)$ are independent. (You have the freedom of choice for the values of the hyperparameters μ, τ^2, n_0 and S_0.)

(a) Obtain the posterior density π for (μ, σ^2), for (μ, σ^{-2}) and for $(\mu, \log \sigma^2)$.

(b) Obtain the posterior mode of $\pi(\mu, \sigma^2)$ using Newton-Raphson and conditional modes iterations.

(c) Calculate $E(\mu|x)$ and $E(\sigma^2|x)$ using the approximating methods below:

  c1. Normal approximation;
  c2. Standard and exponential form Laplace approximation;
  c3. Gaussian quadrature.

(d) Repeat (c) basing your calculations on a reparametrization $h(\sigma^2)$.

(e) Compare the methods about computational aspects and the numeric results obtained and discuss an appropriate choice for the transformation h.

12. Consider the problem of estimating $I = \int t(\theta)\pi(\theta)d\theta$ by

$$\hat{I} = \frac{1}{n}\sum_{i=1}^{n} \frac{t(\theta_i)\pi(\theta_i)}{q(\theta_i)}$$

where $\theta_1, ..., \theta_n$ is a sample from q. Show that

(a) $E_q(\hat{I}) = I$, where E_q denotes expectation with respect to q;

(b) $V_q(\hat{I}) = \sigma^2/n$ where $\sigma^2 = \int [t^2(\theta)\pi^2(\theta)/q(\theta)]d\theta - I^2$;

(c) the optimal choice for q in terms of minimizing σ^2 is $q \propto t \times \pi$.

13. Show that the Monte Carlo ratio estimator of $E[t(\theta)] = I_2/I_1$ given by

$$\hat{I} = \frac{\hat{I}_2}{\hat{I}_1} = \frac{\sum_{i=1}^{n} t(\theta_i)\pi^*(\theta_i)/q(\theta_i)}{\sum_{i=1}^{n} \pi^*(\theta_i)/q(\theta_i)}$$

is asymptotically unbiased and a strongly consistent estimator of I_2/I_1.
Also, derive the approximation

$$Var(\hat{I}) \doteq \frac{Var(\hat{I}_2)}{I_1^2} - 2\frac{I_2}{I_1^3}Cov(\hat{I}_2, \hat{I}_1) + \frac{I_2^2}{I_1^4}Var(\hat{I}_1)$$

using a Taylor series expansion of $\hat{I}_1^{-1}$ around I_1 and propose an estimator for $Var(\hat{I})$.

14. Assume a sample $\theta_1, ..., \theta_n$ from π is available. Show that the estimators below are unbiased and strongly consistent estimators of their corresponding posterior quantities and evaluate the sampling variance of these estimators:

(a) Quantity of interest: $E(\theta)$, estimator: $\bar{\theta} = (1/n)\sum_{i=1}^{n} \theta_i$.

(b) Quantity of interest: $E(\phi)$ where $\phi = t(\theta)$, estimator: $\bar{\phi} = (1/n)\sum_{i=1}^{n} \phi_i$ where $\phi_i = t(\theta_i)$, $i = 1, ..., n$.

(c) Quantity of interest: $p = Pr(\theta \in C)$, estimator: $\hat{p} = (1/n)\sum_{i=1}^{n} I(\theta_i \in C)$.

CHAPTER 4

Markov chains

4.1 Introduction

The presentation of Markov chains in this chapter is far from comprehensive. It tries to describe the main results by combining intuition with probabilistic reasoning. Its presence in this book is an attempt to provide in simple terms the theory governing the iterative simulation techniques used in later chapters. There are many excellent books on or including a detailed and formal treatment of the subject. Among them, the books by Feller (1968), Meyn and Tweedie (1993), Nummelin (1984) and Ross (1996) can be cited. A more statistically oriented approach is given in Guttorp (1995) and a more thorough treatment is given in Revuz (1975).

Markov dependence is a concept attributed to the Russian mathematician Andrei Andreivich Markov that at the start of the 20th century investigated the alternance of vowels and consonants in the poem *Onegin* by Poeshkin. He developed a probabilistic model where successive results depended on all their predecessors only through the immediate predecessor. The model allowed him to obtain good estimates of the relative frequency of vowels in the poem. Almost at the same time the French mathematician Henri Poincaré studied sequences of random variables that were in fact Markov chains.

A Markov chain is a special type of stochastic process, which deals with characterization of sequences of random variables. Special interest is paid to the dynamic and the limiting behaviours of the sequence. A stochastic process can be defined as a collection of random quantities $\{\theta^{(t)} : t \in T\}$ for some set T.

The set $\{\theta^{(t)} : t \in T\}$, is said to be a stochastic process with state space S and index (or parameter) set T. Throughout this book the index set T is taken as countable, defining a discrete time stochastic process. Without loss of generality, it will be assumed to be the set of natural numbers N and will mostly represent the iterations of a simulation scheme. The state space will in general be a subset of R^d representing the support of a parameter vector. For presentation purposes, it will be assumed to be discrete for the first sections of this chapter. Most important results for general Markov chains can be obtained with a discrete state space. Limiting results concerning ergodic and central limit theorems are especially relevant in this book. Later sections consider general state spaces. After the ground work has been done, connections with iterative simulation schemes are outlined and illustrated for one such specific scheme.

4.2 Definition and transition probabilities

In simple terms, a Markov chain is a stochastic process where given the present state, past and future states are independent. This property can be more formally stated through

$$Pr(\theta^{(n+1)} \in A | \theta^{(n)} = x, \theta^{(n-1)} \in A_{n-1}, ..., \theta^{(0)} \in A_0)$$
$$= Pr(\theta^{(n+1)} \in A | \theta^{(n)} = x) \qquad (4.1)$$

for all sets $A_0, ..., A_{n-1}, A \subset S$ and $x \in S$. The Markovian property (4.1) can also be established in the equivalent forms:

1. $E[f(\theta^{(n)}) | \theta^{(m)}, \theta^{(m-1)}, ..., \theta^{(0)}] = E[f(\theta^{(n)}) | \theta^{(m)}]$ for all bounded functions f and $n > m \geq 0$;
2. $Pr(\theta^{(n+1)} = y | \theta^{(n)} = x, \theta^{(n-1)} = x_{n-1}, ..., \theta^{(0)} = x_0) = Pr(\theta^{(n+1)} = y | \theta^{(n)} = x)$ for all $x_0, ..., x_{n-1}, x, y \in S$.

The above form is clearly appropriate only for discrete state spaces. In fact, it is more appropriate than (4.1) in this case and is used as defining Markov chains for the initial sections of this chapter. In general, the probabilities in (4.1) depend on x, A and n. When they do not depend on n, the chain is said to be homogeneous. In this case, a transition function or kernel $P(x, A)$ can be defined as:

1. for all $x \in S$, $P(x, \cdot)$ is a probability distribution over S;
2. for all $A \subset S$, the function $x \mapsto P(x, A)$ can be evaluated.

It is also useful when dealing with discrete state space to identify $P(x, \{y\}) = P(x, y)$. This function is called a transition probability and satisfies:

- $P(x, y) \geq 0, \forall x, y \in S$;
- $\sum_{y \in S} P(x, y) = 1, \forall x \in S$;

as any probability distribution $P(x, \cdot)$ should.

Example 4.1 Random walk

Consider a particle moving independently left and right on the line with successive displacements from its current position governed by a probability function f over the integers and $\theta^{(n)}$ representing its position at instant n, $n \in N$. Initially, $\theta^{(0)}$ is distributed according to some distribution $\pi^{(0)}$. The positions can be related as

$$\theta^{(n)} = \theta^{(n-1)} + w_n = w_1 + w_2 + ... + w_n$$

where the w_i are independent random variables with probability function f. So, $\{\theta^{(n)} : n \in N\}$ is a Markov chain in Z.

The position of the chain at instant $t = n$ is described probabilistically by the distribution of $w_1 + \dots + w_n$.

If $f(1) = p$, $f(-1) = q$ and $f(0) = r$ with $p + q + r = 1$ then the transition probabilities are given by

$$P(x, y) = \begin{cases} p & \text{, if } y = x + 1 \\ q & \text{, if } y = x - 1 \\ r & \text{, if } y = x \\ 0 & \text{, if } y \neq x - 1, x, x + 1 \end{cases}$$

Example 4.2 Branching processes

Consider particles that can generate new particles of the same type. Each of the x particles at generation n gives birth independently to identically distributed numbers of descendants ξ_i, $i = 1, \dots, x$ with discrete distribution F and dies. If $\theta^{(n)}$ represents the number of particles at generation n then it is a Markov chain with state space $S = \{0, 1, 2, \dots\}$ and transition probabilities

$$P(x, y) = Pr \left(\sum_{i=1}^{x} \xi_i = y \right).$$

Note that $P(0, 0) = 1$ and once state 0 is reached, the chain does not leave it. Such states are called absorbing states.

Example 4.3 Ehrenfest model

Consider a total of r balls distributed in two urns with x balls in the first urn and $r - x$ in the second urn. Take one of the r balls at random and put it in the other urn. Repeat the random selection process independently and indefinitely. This procedure was used by Ehrenfest to model the exchange of molecules between two containers. If $X^{(n)}$ represents the number of balls in the first urn after n exchanges then $\{X^{(n)} : n \in N\}$ is a Markov chain with state space $S = \{0, 1, 2, \dots, r\}$ and transition probabilities

$$P(x, y) = \begin{cases} x/d & \text{, if } y = x + 1 \\ 1 - x/d & \text{, if } y = x - 1 \\ 0 & \text{, if } |y - x| \neq 1 \end{cases}$$

Example 4.4 Birth and death processes

Consider a Markov chain that from the state x can only move in the next step to one of the neighbouring states $x - 1$, representing a death, x or $x + 1$, representing a birth. The transition probabilities are given by

$$P(x, y) = \begin{cases} p_x & \text{, if } y = x + 1 \\ q_x & \text{, if } y = x - 1 \\ r_x & \text{, if } y = x \\ 0 & \text{, if } |y - x| > 1 \end{cases}$$

where p_x, q_x and r_x are non-negative with $p_x + q_x + r_x = 1$ and $q_0 = 0$.
The Ehrenfest model is a special case of birth and death processes.

In the case of discrete state spaces $S = \{x_1, x_2, ...\}$, a transition matrix
P with (i, j)th element given by $P(x_i, x_j)$ can be defined. If S is finite with
r elements, the transition matrix P is given by

$$P = \begin{pmatrix} P(x_1, x_1) & \cdots & P(x_1, x_r) \\ \vdots & & \vdots \\ P(x_r, x_1) & \cdots & P(x_r, x_r) \end{pmatrix}$$

Transition matrices have all lines summing to one. Such matrices are called
stochastic and have a few interesting properties. For instance, at least one
eigenvalue of a stochastic matrix equals one and the product of stochastic
matrices always produces a stochastic matrix. Of course, countable state
spaces will lead to an infinite number of eigenvalues.

Transition probabilities from state x to state y over m steps, denoted
by $P^m(x, y)$, is given by the probability of a chain moving from state x to
state y in exactly m steps. It can be obtained for $m \geq 2$ as

$$P^m(x, y) = Pr(\theta^{(m)} = y | \theta^{(0)} = x)$$

$$= \sum_{x_1} \cdots \sum_{x_{m-1}} Pr(\theta^{(m)} = y, \theta^{(m-1)} = x_{m-1}, ..., \theta^{(1)} = x_1 | \theta^{(0)} = x)$$

$$= \sum_{x_1} \cdots \sum_{x_{m-1}} Pr(\theta^{(m)} = y | \theta^{(m-1)} = x_{m-1}) ... Pr(\theta^{(1)} = x_1 | \theta^{(0)} = x)$$

$$= \sum_{x_1} \cdots \sum_{x_{m-1}} P(x, x_1) P(x_1, x_2) \cdots P(x_{m-1}, y)$$

where the second equality is due to the Markovian property of the process.
The last equality means that the matrix containing elements $P^m(x, y)$ is
also a stochastic matrix and is given by P^m obtained by the matrix product
of the transition matrix P m times. Also, for completeness, $P^1(x, y) = P(x, y)$ and $P^0(x, y) = I(x = y)$. The above derivation can be used to
established that

$$P^{n+m}(x, y) = \sum_z Pr(\theta^{(n+m)} = y | \theta^{(n)} = z, \theta^{(0)} = x) Pr(\theta^{(n)} = z | \theta^{(0)} = x)$$

$$= \sum_z P^n(x, z) P^m(z, y). \tag{4.2}$$

Equations (4.2) are usually called Chapman-Kolmogorov equations. All
summations are with respect to the elements of the state space S and results
are valid for any stage of the chain due to the assumed homogeneity. Higher
transition matrices can be formed with these higher transition probabilities

and it can be shown that they satisfy the relation $P^{n+m} = P^n P^m$ and, in particular, $P^{n+1} = P^n P$.

The marginal distribution of the nth stage can be defined by the row vector $\pi^{(n)}$ with components $\pi^{(n)}(x_i)$, for all $x_i \in S$. For finite state spaces, this is a r-dimensional vector

$$\pi^{(n)} = (\pi^{(n)}(x_1), \cdots, \pi^{(n)}(x_r))$$

When $n = 0$, this is the initial distribution of the chain. Then,

$$
\begin{aligned}
\pi^{(n)}(y) &= Pr(\theta^{(n)} = y) \\
&= \sum_{x \in S} Pr(\theta^{(n)} = y | \theta^{(0)} = x) Pr(\theta^{(0)} = x) \\
&= \sum_{x \in S} P^n(x, y) \pi^{(0)}(x).
\end{aligned}
$$

The above equation can be written in matrix notation as $\pi^{(n)} = \pi^{(0)} P^n$. Also, since the same is valid for $n - 1$, $\pi^{(n)} = \pi^{(0)} P^{n-1} P = \pi^{(n-1)} P$.

The probability of any event $A \subset S$ for a Markov chain starting at x is denoted by $Pr_x(A)$. The hitting time of A is defined as $T_A = \min\{n \geq 1 : \theta^{(n)} \in A\}$ if $\theta^{(n)} \in A$ for some $n > 0$. Otherwise, $T_A = \infty$. If $A = \{a\}$, the notation $T_{\{a\}} = T_a$ is used.

Example 4.5 Consider $\{\theta^{(n)} : n \geq 0\}$, a Markov chain in $S = \{0, 1\}$ with initial distribution $\pi^{(0)}$ given by $\pi^{(0)} = (\pi^{(0)}(0), \pi^{(0)}(1))$ and transition matrix P given by $P = \begin{pmatrix} 1-p & p \\ q & 1-q \end{pmatrix}$.

Using the relation $Pr(\theta^{(n)} = 0) = \sum_{j \in S} Pr(\theta^{(n)} = 0, \theta^{(n-1)} = j)$ for $n \geq 1$ and the Markovian property of the chain,

$$
\begin{aligned}
Pr(\theta^{(n)} = 0) &= (1-p) Pr(\theta^{(n-1)} = 0) + q Pr(\theta^{(n-1)} = 1) \\
&= (1-p-q)^n \pi^{(0)}(0) + q \sum_{k=0}^{n-1} (1-p-q)^k
\end{aligned}
$$

and $Pr(\theta^{(n)} = 1) = 1 - Pr(\theta^{(n)} = 0)$. If $p = q = 0$, the chain never moves with probability one and $Pr(\theta^{(n)} = 0) = \pi_0(0)$ and $Pr(\theta^{(n)} = 1) = \pi_0(1)$. If $p + q > 0$,

$$Pr(\theta^{(n)} = 0) = \frac{q}{p+q} + (1-p-q)^n \left(\pi^{(0)}(0) - \frac{q}{p+q} \right)$$

and $Pr(\theta^{(n)} = 1) = 1 - Pr(\theta^{(n)} = 0)$. Note that if $\pi^{(0)} = (q, p)/(p+q)$, then $Pr(\theta^{(n)} = 0) = q/(p+q)$, for all n, and the distributions at all steps are the same as the initial distribution.

If, in addition, $p + q < 2$ then

$$\lim_{n \to \infty} Pr(\theta^{(n)} = 0) = \frac{q}{p+q} \quad \text{and} \quad \lim_{n \to \infty} Pr(\theta^{(n)} = 1) = \frac{p}{p+q}$$

and the initial distribution is obtained now as a limiting distribution of the chain. The value of the higher transition matrix P^n can be found via $Pr(\theta^{(n)} = 0) = \pi^{(n)}(0) = \sum_x P^n(x, 0)\pi^{(0)}(x)$, for all x. Taking $\pi^{(0)}(0) = 1$ gives

$$P^n(0, 0) = \frac{q}{p+q} + (1 - p - q)^n \frac{p}{p+q}$$

Analogously for $P^n(0, 1)$, $P^n(1, 0)$ and $P^n(1, 1)$ gives

$$P^n = \frac{1}{p+q} \begin{pmatrix} q & p \\ q & p \end{pmatrix} + \frac{(1 - p - q)^n}{p+q} \begin{pmatrix} p & -p \\ -q & q \end{pmatrix} \tag{4.3}$$

Finally $Pr_0(T_x = n)$, $x = 0, 1$ can be found by

$$\begin{aligned}
Pr_0(T_0 = n) &= Pr_0(\theta^{(n)} = 0, \theta^{(j)} \neq 0, 1 \leq j \leq n - 1) \\
&= P(1, 0)P(1, 1)^{n-2}P(1, 0) \\
&= q(1 - q)^{n-2}p.
\end{aligned}$$

Analogously, $Pr_0(T_1 = n) = p(1 - p)^{n-1}$.

It is important to distinguish between $P^n(0, 0) = Pr_0(\theta^{(n)} = 0)$, the probability of the chain, starting from state 0, hitting state 0 in n steps, and $Pr_0(T_0 = n) = Pr_0(\theta^{(n)} = 0, \theta^{(j)} \neq 0, j = 1, 2, ..., n - 1)$, the probability of the chain, starting from state 0, hitting state 0 in n steps for the first time.

4.3 Decomposition of the state space

A few quantities of interest will be defined. They will be important in the classification of states of a Markov chain with state space S and transition matrix P.

- The probability of the chain starting from state x hitting state y at any posterior step is

$$\rho_{xy} = Pr_x(T_y < \infty)$$

- The number of visits of a chain to a state y is

$$N(y) = \#\{n > 0 : \theta^{(n)} = y\} = \sum_{n=1}^{\infty} I(\theta^{(n)} = y)$$

It can be shown that $E(T_y \mid \theta^{(0)} = x) = \sum_{n=0}^{\infty} Pr_x(T_y > n)$ and $E(N(y) \mid \theta^{(0)} = x) = \sum_{n=1}^{\infty} P^n(x, y)$.

A state $y \in S$ is said to be recurrent if the Markov chain, starting in y, returns to y with probability 1 ($\rho_{yy} = 1$) and is said to be transient if it has positive probability of not returning to y ($\rho_{yy} < 1$). An absorbing state $y \in S$ is recurrent because $Pr_y(T_y = 1) = Pr_y(\theta^{(1)} = y) = P(y, y) = 1$

and therefore $\rho_{yy} = 1$. If a Markov chain starts at a recurrent state y, the hitting (or return, in this case) time of y, T_y, is a finite random quantity whose mean μ_y can be evaluated. If this mean is finite, the state y is said to be positive recurrent and otherwise the state is said to be null recurrent. Positive recurrence is a very important property for establishing limiting results, as will be seen in the next section.

An important result describing analytically the difference between a recurrent and a transient state is that

- if $y \in S$ is a transient state then, for all $x \in S$,

$$Pr_x(N(y) < \infty) = 1 \text{ and } E[N(y) \mid \theta^{(0)} = x] = \frac{\rho_{xy}}{1 - \rho_{yy}} < \infty$$

- if $y \in S$ is a recurrent state then

$$Pr_y(N(y) = \infty) = 1 \text{ and } E[N(y) \mid \theta^{(0)} = y] = \infty$$

So, recurrent states are infinitely often (i.o.) visited with probability one. The expected number of visits is finite if the state is transient.

It is interesting to investigate possible decompositions of S in subsets of recurrent and transient states. From this decomposition, probabilities of the chain hitting a given set of states can be evaluated. For states x and y in S, $x \neq y$, x is said to hit y, denoted $x \to y$, if $\rho_{xy} > 0$. A set $C \subseteq S$ is said to be closed if

$$\rho_{xy} = 0 \text{ for } x \in C \text{ and } y \notin C.$$

In obvious nomenclature, it is said to be irreducible if $x \to y$ for every pair $x, y \in C$. A chain is said to be irreducible if S is irreducible. It is not difficult to show that the condition $\rho_{xy} > 0$ is equivalent to $P^n(x, y) > 0$ for some $n > 0$. This can be used to show that if $x \in S$ is recurrent and $x \to y$ then y is also recurrent. In this case, $y \to x$ and one can write $x \leftrightarrow y$ when $x \to y$ and $y \to x$. In other words, recurrence defines an equivalence class with respect to the $\leftrightarrow$ operation. Also, $\rho_{xy} = \rho_{yx} = 1$. In fact, a stronger result is valid: null recurrence and positive recurrence also define equivalence classes (Guttorp, 1995). If $C \subseteq S$ is a closed, finite, irreducible set of states then all states of C are recurrent.

Example 4.6 Consider the transition matrices over $S = \{0, 1, ..., r\}$ below. For each one of them, S can be decomposed into S_R and S_T, the sets of recurrent and transient states respectively. The symbols $+$ and $-$ associated with the pair (x, y) denote $x \to y$ and $x \not\to y$, respectively.

a) $\quad P = \begin{pmatrix} 1/2 & 1/2 & 0 \\ 1/2 & 1/4 & 1/4 \\ 0 & 1/3 & 2/3 \end{pmatrix}$ $\qquad \begin{pmatrix} + & + & + \\ + & + & + \\ + & + & + \end{pmatrix}$

Therefore, S is closed and the chain is irreducible. To show that, for instance, $0 \to 2$, it suffices to verify that $P^2(0,2) > 0$. So, $S = S_R$.

$$
b)\ P = \begin{pmatrix}
1 & 0 & 0 & 0 & 0 & 0 \\
1/4 & 1/2 & 1/4 & 0 & 0 & 0 \\
0 & 1/5 & 2/5 & 1/5 & 0 & 1/5 \\
0 & 0 & 0 & 1/6 & 1/3 & 1/2 \\
0 & 0 & 0 & 1/2 & 0 & 1/2 \\
0 & 0 & 0 & 1/4 & 0 & 3/4
\end{pmatrix}
\begin{pmatrix}
+ & - & - & - & - & - \\
+ & + & + & + & + & + \\
+ & + & + & + & + & + \\
- & - & - & + & + & + \\
- & - & - & + & + & + \\
- & - & - & + & + & +
\end{pmatrix}
$$

In this case, $S_R = \{0\} \cup \{3,4,5\}$ and $S_T = \{1,2\}$.

If the set of recurrent states S_R is not empty then it can be written as a (finite or countable) union of disjoint, closed and irreducible sets. Whenever the chain hits a closed, irreducible set C of recurrent states, it stays in C forever and with probability one it visits all its elements i.o. It is interesting to compute ρ_{xy} for x transient and y recurrent. As y belongs to an irreducible set C, $\rho_{xy} = \rho_C(x) = Pr_x(T_C < \infty)$ is called the absorption probability. If the chain starts at $x \in S_T$, it can hit C in the first step or remain in S_T in the first step and hit C at a later step. So,

$$
\rho_C(x) = \sum_{y \in C} P(x,y) + \sum_{y \in S_T} P(x,y)\rho_C(y), \quad x \in S_T. \tag{4.4}
$$

The uniqueness of solutions of this system is guaranteed if $Pr_x(T_{S_R} < \infty) = 1$, for $x \in S_T$. In the case of a finite S_T it is automatically valid as transient states are only finitely visited.

Example 4.6 (continued) Let $S_R = C_1 \cup C_2$ where $C_1 = \{0\}$ and $C_2 = \{3,4,5\}$. The absorption probabilities $\rho_{10} = \rho_{C_1}(1)$ and $\rho_{20} = \rho_{C_1}(2)$ are obtained from Equation (4.4) as

$$
\rho_{10} = \frac{1}{4} + \frac{1}{2}\rho_{10} + \frac{1}{4}\rho_{20} \text{ and } \rho_{20} = 0 + \frac{1}{5}\rho_{10} + \frac{2}{5}\rho_{20}
$$

Solving for ρ_{10} and ρ_{20} gives $\rho_{10} = 3/5$ and $\rho_{20} = 1/5$.

Proceeding analogously for C_2 gives $\rho_{C_2}(1) = 2/5$ and $\rho_{C_2}(2) = 4/5$. As S_T is finite, the absorption probabilities $\rho_{C_2}(x)$ can be evaluated for $x \in S_T$ through $\sum_i \rho_{C_i}(x) = 1$.

Example 4.4 (continued) Irreducible chains are obtained when $p_x > 0$ for $x \geq 0$ and $q_x > 0$ for $x > 0$. It is possible to determine if a state y is recurrent or transient even for an infinite state space by studying the convergence of the series $\sum_{y=0}^{\infty} \gamma_y$ where

$$
\gamma_y = \begin{cases}
1 & , \text{if } y = 0 \\
\frac{q_1 \ldots q_y}{p_1 \ldots p_y} & , \text{if } y > 0
\end{cases}
$$

If the sum diverges, the chain is recurrent. Otherwise, the chain is transient. If S is finite and 0 is an absorbing state, the absortion probability is

$$\rho_{\{0\}}(x) = \rho_{x0} = \frac{\sum_{y=x}^{d-1} \gamma_y}{\sum_{y=0}^{d-1} \gamma_y} \quad , \quad x = 1, ..., d-1.$$

Details of these calculations are given, for example, in Hoel, Port and Stone (1972).

4.4 Stationary distributions

A fundamental problem for Markov chains in the context of simulation is the study of the asymptotic behaviour of the chain as the number of steps or iterations $n \to \infty$. A key concept is that of a stationary distribution π. A distribution π is said to be a stationary distribution of a chain with transition probabilities $P(x, y)$ if

$$\sum_{x \in S} \pi(x) P(x, y) = \pi(y), \quad \forall y \in S. \tag{4.5}$$

Equation (4.5) can be written in matrix notation as $\pi = \pi P$. The reason of the name is clear from the above equation. If the marginal distribution at any given step n is π then the distribution at the next step is $\pi P = \pi$. Once the chain reaches a stage where π is the distribution of the chain, the chain retains this distribution for all subsequent stages. This distribution is also known as the invariant or equilibrium distribution for similar interpretations.

It will be shown below that if the stationary distribution π exists and $\lim_{n \to \infty} P^n(x, y) = \pi(y)$ then, independently of the initial distribution of the chain, $\pi^{(n)}$ will approach π, as $n \to \infty$. In this sense, the distribution is also referred to as the limiting distribution.

Example 4.5 (continued) The stationary distribution π is the solution of the system $\pi P = \pi$ that gives equations

$$\pi(0) P(0, y) + \pi(1) P(1, y) = \pi(y), \quad y = 0, 1.$$

The solution is $\pi = (q, p)/(p + q)$, a distribution that was shown to be invariant for the stages of the chain.

Also, provided $p + q < 2$, $\lim_{n \to \infty} P^n = \frac{1}{p+q} \begin{pmatrix} q & p \\ q & p \end{pmatrix}$ from (4.3) and

the distribution of $\theta^{(n)}$ converges to π at an exponential rate.

The case $p + q = 2$ still produces a stationary distribution π but this does not provide a unique limiting distribution as $Pr(\theta^{(n)} = 0) = (1/2) + (-1)^n [\pi^{(0)} - (1/2)]$, for all n. This case is somehow different because the states are always alternating through the stages. It has a periodic nature that will be addressed in the next section.

Example 4.7 Gibbs sampler (Geman and Geman, 1984)

This example provides a very simple special case of the Gibbs sampler and was considered by Casella and George (1992). The complete form of the Gibbs sampler will be left for the next chapter. In this special case, the state space is $S = \{0, 1\}^2$ and define a probability distribution π over S as

$$
\begin{array}{c|cc}
 & \multicolumn{2}{c}{\theta_2} \\
 & 0 & 1 \\
\hline
\theta_1 \quad 0 & \pi_{00} & \pi_{01} \\
1 & \pi_{10} & \pi_{11} \\
\end{array}
$$

The probability vector π contains the above probabilities in any fixed order, say $(\pi_{00}, \pi_{01}, \pi_{10}, \pi_{11})$.

The chain now consists of a bidimensional vector $\theta^{(n)} = (\theta_1^{(n)}, \theta_2^{(n)})$. Although this introduces some novelties in the presentation they can easily be removed by considering a scalar chain $\psi^{(n)}$ that assumes values that are in correspondence with the $\theta^{(n)}$ chain, e.g. $\psi^{(n)} = 10\theta_1^{(n)} + \theta_2^{(n)}$. This is always possible for discrete state spaces and from now on no distinction will be made between scalar and vector chains.

Consider the following transition probabilities:

- For the first component θ_1, the transition probabilities are given by the conditional distribution π_1 of $\theta_1 | \theta_2 = j$,

$$
\pi_1(0|j) = \frac{\pi_{0j}}{\pi_{+j}} \text{ and } \pi_1(1|j) = \frac{\pi_{1j}}{\pi_{+j}}
$$

where $\pi_{+j} = \pi_{0j} + \pi_{1j}$, $j = 0, 1$.

- For the second component θ_2, the transition probabilities are given by the conditional distribution π_2 of $\theta_2 | \theta_1 = i$,

$$
\pi_2(0|i) = \frac{\pi_{i0}}{\pi_{i+}} \text{ and } \pi_2(1|i) = \frac{\pi_{i1}}{\pi_{i+}}
$$

where $\pi_{i+} = \pi_{i0} + \pi_{i1}$, $i = 0, 1$.

The overall transition probability of the chain is

$$
\begin{aligned}
P((i, j), (k, l)) &= Pr(\theta^{(n)} = (k, l) | \theta^{(n-1)} = (i, j)) \\
&= Pr(\theta_2^{(n)} = l | \theta_1^{(n)} = k) \, Pr(\theta_1^{(n)} = k | \theta_2^{(n-1)} = j) \\
&= \frac{\pi_{kl}}{\pi_{k+}} \frac{\pi_{kj}}{\pi_{+j}}
\end{aligned}
$$

for $(i, j), (k, l) \in S$. Thus, a 4×4 transition matrix P can be formed.

It is evident from the above transitions that $(\theta^{(n)})_{n \geq 0}$ forms a Markov chain as transition probabilities only depend on the present value (k, l) to predict the future value (i, j). It is straightforward to ascertain that π is the stationary distribution of the chain. If all elements of π are positive, it is also a limiting distribution. The interesting message is that chains formed by the superposition of conditional distributions have a stationary distribution given by the joint distribution.

The results can be extended in the same way for cases when θ_1 can take m_1 values and θ_2 can take m_2 values. They can similarly be extended to cases where the θ consist of d components and each of them can take m_i values, $i = 1, ..., d$. The Gibbs sampler is a simulation scheme that will be applied in the next chapter for general (continuous, discrete and mixed) d-dimensional state spaces.

Example 4.4 (continued) For irreducible birth and death chains, the system of equations $\sum_{x \in S} \pi(x) P(x, y) = \pi(y)$, $y \in S$, is $\pi(x) = p_{x-1}\pi(x - 1) + r_x\pi(x) + q_{x+1}\pi(x + 1)$ for $x > 0$ and $\pi(0) = r_0\pi(0) + q_1\pi(1)$ for $x = 0$. Using the fact that $r_x = 1 - p_x - q_x$, these equations reduce to

$$\pi(x + 1) = \frac{p_x}{q_{x+1}} \pi(x), \quad x \geq 0.$$

Defining

$$\pi_x = \begin{cases} 1 & \text{, if } x = 0 \\ \frac{p_0 p_1 \cdots p_{x-1}}{q_1 q_2 \cdots q_x} & \text{, if } x \geq 1 \end{cases}$$

gives that, if $\sum_x \pi_x$ converges, the birth and death chain has stationary distribution

$$\pi(x) = \frac{\pi_x}{\sum_{y=0}^{\infty} \pi_y}, \quad x \geq 0.$$

In the case of an Ehrenfest model where $P(x, x) = r_x = 0$ for all x, $P^n(x, x) = 0$ for odd n. Hence, the chain can only return to the same state after an even number of transitions. In this case, there is no limiting distribution, as will be shown below.

The existence and uniqueness of stationary distributions can be studied through weaker results. Let $N_n(y)$ be the number of visits to state y in n steps and define $G_n(x, y) = E_x[N_n(y)]$, the average number of visits of the chain to state y and $m_y = E_y(T_y)$, the average return time to state y. Then, $G_n(x, y) = \sum_{k=1}^{n} P^k(x, y)$ and $\lim_{n \to \infty} G_n(x, y)/n$ provides the limiting occupation of state y in a chain observed for an infinitely long number of steps. It can be shown that

- If $y \in S$ is transient then $\lim_{n \to \infty} \frac{N_n(y)}{n} = 0$, with probability 1, and $\lim_{n \to \infty} \frac{G_n(x,y)}{n} = 0$ for all $x \in S$.

- If $y \in S$ is recurrent then $\lim_{n \to \infty} \frac{N_n(y)}{n} = \frac{I(T_y < \infty)}{m_y}$, with probability 1, and $\lim_{n \to \infty} \frac{G_n(x,y)}{n} = \frac{\rho_{xy}}{m_y}$, for all $x \in S$.

It follows that if π is a stationary distribution then $\pi(x) = 0$, if x is transient or null recurrent ($m_y = \infty$), and $\pi(x) = \frac{1}{m_x}$, if x is positive recurrent. Intuitively, the stationary probability of any state is given by the frequency of visits to the state. As the sets S_{Rp} and S_{Rn} of positive recurrent and null recurrent states are closed if S is finite, then $S_{Rn} = \phi$, the empty set. Therefore, an irreducible Markov chain is positive recurrent if and only if it possesses a stationary distribution π such that

$$\lim_{n \to \infty} \frac{\sum_{k=1}^{n} P^k(x,y)}{n} = \lim_{n \to \infty} \frac{G_n(x,y)}{n} = \pi(y).$$

If the chain is null recurrent then Equation (4.5) is still valid but the $\pi(y)$s do not possess a finite sum and therefore do not constitute a proper distribution.

The analysis of the limiting behaviour of P^n is more delicate for technical reasons and is considered in the next section. It is clear that if $y \in S_T$ or $y \in S_{Rn}$ then $\lim_{n \to \infty} P^n(x,y) = 0$, $\forall x \in S$. If $y \in S_{Rp}$, it has only been obtained that $\lim_{n \to \infty} P^n(x,y)$ must equal ρ_{xy}/m_y.

4.5 Limiting theorems

There are situations where stationary distributions are available but limiting distributions are not (Example 4.5). In order to establish limiting results, one characterization of states still absent must be introduced. This is the notion of periodicity.

The period of a state x, denoted by d_x, is the largest common divisor of the set

$$\{n \geq 1 : P^n(x,x) > 0\}.$$

It is obvious that $P(x,x) > 0$ implies that $d_x = 1$ and that if $x \leftrightarrow y$ then $d_x = d_y$. Therefore, the states of an irreducible chain have the same period. A state x is aperiodic if $d_x = 1$ and if in addition the state is positive recurrent, the state is said to be ergodic. A chain is periodic with period d if all its states are periodic with period $d > 1$ and aperiodic if all its states are aperiodic. Finally, a chain is ergodic if all its states are ergodic.

Although aperiodicity is superfluous for the existence of an equilibrium distribution, it is a condition needed to establish convergence of the transition probabilities. Let $(\theta^{(n)})_{n \geq 0}$ be an irreducible, positive recurrent chain with stationary distribution π and $\|\xi_1 - \xi_2\| = \sup_{A \subset S} |\xi_1(A) - \xi_2(A)|$ be the total variation distance between two distributions ξ_1 and ξ_2.

- If the chain is aperiodic then $\lim_{n \to \infty} P^n(x,y) = \pi(y)$ for all $x, y \in S$. In fact, irreducibility and ergodicity of the chain are equivalent to $\lim_{n \to \infty} \|P^n(x, \cdot) - \pi(\cdot)\| = 0$, for all $x \in S$ (Nummelin, 1984).

- If the chain is periodic with period d then, for every $x, y \in S$, there is an integer r, $0 \leq r < d$ such that $P^n(x,y) = 0$ unless $n = md + r$ for some $m \in N$ and $\lim_{m \to \infty} P^{md+r}(x,y) = d\pi(y)$.

Example 4.3 (continued) If $y - x$ is even then $P^{2m+1}(x, y) = 0$, $m \geq 0$ and $\lim_{m \to \infty} P^{2m}(x, y) = 2\pi(y)$, and if $y - x$ is odd then $P^{2m}(x, y) = 0$, $m \geq 0$ and $\lim_{m \to \infty} P^{2m+1}(x, y) = 2\pi(y)$. The transition matrix P is periodic with period $d = 2$.

Let $S = \{0, 1, 2, 3\}$ $(r = 3)$. From equation (4.5),

$$\pi = \left(\frac{1}{8}, \frac{3}{8}, \frac{3}{8}, \frac{1}{8} \right)$$

and the limiting distributions are obtained from

$$\lim_{n \to \infty} P^n = \begin{pmatrix} \frac{1}{4} & 0 & \frac{3}{4} & 0 \\ 0 & \frac{3}{4} & 0 & \frac{1}{4} \\ \frac{1}{4} & 0 & \frac{3}{4} & 0 \\ 0 & \frac{3}{4} & 0 & \frac{1}{4} \end{pmatrix} \quad \text{for even } n \text{ and}$$

$$\lim_{n \to \infty} P^n = \begin{pmatrix} 0 & \frac{3}{4} & 0 & \frac{1}{4} \\ \frac{1}{4} & 0 & \frac{3}{4} & 0 \\ 0 & \frac{3}{4} & 0 & \frac{1}{4} \\ \frac{1}{4} & 0 & \frac{3}{4} & 0 \end{pmatrix} \quad \text{for odd } n$$

Once ergodicity of the chain is established, important limiting theorems can be stated. The first and most important one is the ergodic theorem. The ergodic average of a real-valued function $t(\theta)$ is the average $\bar{t}_n = (1/n) \sum_{i=1}^{n} t(\theta^{(i)})$.* If the chain is ergodic and $E_\pi[t(\theta)] < \infty$ for the unique limiting distribution π then

$$\bar{t}_n \to E_\pi[t(\theta)] \text{ as } n \to \infty, \text{ with probability } 1 \tag{4.6}$$

This result is a Markov chain equivalent of the law of large numbers (3.8). It states that averages of chain values also provide strongly consistent estimates of parameters of the limiting distribution π despite their dependence. If $t(\theta) = I(\theta = x)$ then the ergodic averages are simply counting the relative frequency of values of xs in realizations of the chain. By the ergodic theorem, this relative frequency converges almost surely to $\pi(x) = 1/m_x$, the average frequency of visits to state x.

Just as there is an equivalent of the law of large numbers for Markov chains, there are also versions of the central limit theorem (3.7) for Markov chains. Many forms are available depending on further conditions on the chain. A chain is said to be geometrically ergodic if there is a constant $0 \leq \lambda < 1$ and a real, integrable function $M(x)$ such that

$$\|P^n(x, \cdot) - \pi(\cdot)\| \leq M(x)\lambda^n \tag{4.7}$$

---

* The average could have included the term corresponding to the initial step and all limiting results would still follow. In the sequel, it will be assumed that chains start at step 1.

for all $x \in S$. If the function M does not depend on x, the ergodicity is uniform. Uniform ergodicity implies geometric ergodicity which implies ergodicity (Tierney, 1994). The smallest λ satisfying (4.7) is called the rate of convergence. Roberts (1996) provides a brief discussion showing that this rate is bounded by the second largest eigenvalue of P. Of course, a geometric rate is desirable for a fast convergence to the limiting distribution. However, this speed may be offset by a very large value of $M(x)$ which may slow down convergence considerably (Polson, 1996). Also, the rate of convergence can be arbitrarily close to 1 (Example 5.5).

Before stating central limit theorems, define the autocovariance of lag k $(k \geq 0)$ of the chain $t^{(n)} = t(\theta^{(n)})$ as $\gamma_k = Cov_\pi(t^{(n)}, t^{(n+k)})$, the variance of $t^{(n)}$ as $\sigma^2 = \gamma_0$, the autocorrelation of lag k as $\rho_k = \gamma_k/\sigma^2$ and $\tau_n^2/n = Var_\pi(\bar{t}_n)$. It can be shown that

$$\tau_n^2 = \sigma^2 \left(1 + 2 \sum_{k=1}^{n-1} \frac{n-k}{n} \rho_k \right)$$

and that $\tau_n^2 \to \tau^2$ as $n \to \infty$ where

$$\tau^2 = \sigma^2 \left(1 + 2 \sum_{k=1}^{\infty} \rho_k \right) \tag{4.8}$$

if the series of autocorrelation is summable.

It is important to distinguish between $\sigma^2 = Var_\pi[t(\theta)]$, the variance of $t(\theta)$ under the limiting distribution π and τ^2, the limiting sampling variance of $\sqrt{n}\,\bar{t}$. Note that under independent sampling they are both given by σ^2. They are both variability measures but the first one is a characteristic of the limiting distribution π whereas the second is the uncertainty of the averaging procedure.

If a chain is uniformly (geometrically) ergodic and $t^2(\theta)$ $(t^{2+\epsilon}(\theta))$ is integrable with respect to π (for some $\epsilon > 0$) then

$$\sqrt{n}\,\frac{\bar{t}_n - E_\pi[t(\theta)]}{\tau} \to N(0,1) \text{ in distribution} \tag{4.9}$$

as $n \to \infty$ (Tierney, 1996). Other versions of the central limit theorem may be found in Tierney (1994) and the subsequent discussion by Chan and Geyer (1994). Just as (4.6) provides theoretical support for the use of ergodic averages as estimates, Equation (4.9) provides support for evaluation of approximate confidence intervals. These will require estimation of the unknown quantity τ^2, a subject that will be treated in section 4.8.

4.6 Reversible chains

Let $(\theta^{(n)})_{n \geq 0}$ be an homogeneous Markov chain with transition probabilities $P(x, y)$ and stationary distribution π. Assume that one wishes to study the sequence of states $\theta^{(n)}, \theta^{(n-1)}, \dots$ in reversed order. It can be shown that this sequence satisfies

$$Pr(\theta^{(n)} = y \mid \theta^{(n+1)} = x, \theta^{(n+2)} = x_2, \dots) = Pr(\theta^{(n)} = y \mid \theta^{(n+1)} = x)$$

and therefore defines a Markov chain. The transition probabilities $P_n^*(x, y)$ are defined by

$$
\begin{aligned}
P_n^*(x, y) &= Pr(\theta^{(n)} = y \mid \theta^{(n+1)} = x) \\
&= \frac{Pr(\theta^{(n+1)} = x \mid \theta^{(n)} = y) Pr(\theta^{(n)} = y)}{Pr(\theta^{(n+1)} = x)} \\
&= \frac{\pi^{(n)}(y) P(y, x)}{\pi^{(n+1)}(x)}
\end{aligned}
$$

and in general the chain is not homogeneous. If $n \to \infty$ or alternatively, $\pi^{(0)} = \pi$, then $P_n^*(x, y) = P^*(x, y) = \pi(y) P(y, x)/\pi(x)$ and the chain becomes homogeneous. If $P^*(x, y) = P(x, y)$ for all x and $y \in S$, the time reversed Markov chain has the same transition probabilities as the original Markov chain. Markov chains with such a property are said to be reversible and the reversibility condition is usually written as

$$\pi(x) P(x, y) = \pi(y) P(y, x), \text{ for all } x, y \in S \qquad (4.10)$$

It can be interpreted as saying that the rate at which the system moves from x to y when in equilibrium, $\pi(x) P(x, y)$, is the same as the rate at which it moves from x to y, $\pi(y) P(y, x)$. For that reason, (4.10) is sometimes referred to as the detailed balance equation (Guttorp, 1995); balance because it equates the rates of moves through states and detailed because it does it for every possible pair of states.

Example 4.4 (continued) The stationary distribution has already been obtained. There are only 4 possibilities for a pair $(x, y) \in S$:

- if $|x - y| > 2$, then $P(x, y) = P(y, x) = 0$ and (4.10) is satisfied;
- if $x = y$, then (4.10) is trivially satisfied;
- if $x = y - 1$ then $P(x, y) = q_x$ and $P(y, x) = p_{x-1}$. It has also been shown that $\pi(x) = \pi(x - 1) p_{x-1}/q_x$ for $x > 0$, which again implies that (4.10) is satisfied;
- if $x = y + 1$ then $P(x, y) = p_x$ and $P(y, x) = q_{x-1}$. The same relations of the previous case hold and (4.10) is satisfied.

In conclusion, irreducible birth and death chains are reversible.

Reversible chains are useful because if there is a distribution π satisfying (4.10) for an irreducible chain, then the chain is positive recurrent, reversible with stationary distribution π. This is easily obtained by summing over y both sides of (4.10) to give (4.5). Construction of Markov chains with a given stationary distribution π reduces to finding transition probabilities $P(x,y)$ satisfying (4.10). This is always possible, as the next example shows.

Example 4.8 Metropolis algorithm (Metropolis et al., 1953)
Consider a given distribution p_x, $x \in S$ with $\sum_x p_x = 1$ where the state space S can be a subset of the line or even a d-dimensional subset of R^d. The problem posed and solved by Metropolis et al. (1953) was how to construct a Markov chain with stationary distribution π such that $\pi(x) = p_x$, $x \in S$. Let Q be any irreducible transition matrix on S satisfying the symmetry condition $Q(x,y) = Q(y,x)$, for $x,y \in S$.

Define a Markov chain $(\theta^{(n)})_{n \geq 0}$ as having transition from x to y proposed according to the probabilities $Q(x,y)$. This proposed value for $\theta^{(n+1)}$ is accepted with probability $\min\{1, p_y/p_x\}$ and rejected otherwise, leaving the chain in state x.

The transition probabilities $P(x,y)$ of the above chain $(\theta^{(n)})_{n \geq 0}$ are

$$\begin{aligned} P(x,y) &= Pr(\theta^{(n+1)} = y, TA|\theta^{(n)} = x) \\ &= Pr(\theta^{(n+1)} = y|\theta^{(n)} = x)Pr(TA) \\ &= Q(x,y)\min\{1, p_y/p_x\} \end{aligned}$$

for $y \neq x$ and TA denotes the event [transition is accepted]. If $y = x$, then

$$\begin{aligned} P(x,x) &= Pr(\theta^{(n+1)} = x, TA|\theta^{(n)} = x) + Pr(\theta^{(n+1)} \neq x, \bar{TA}|\theta^{(n)} = x) \\ &= Pr(\theta^{(n+1)} = x|\theta^{(n)} = x)Pr(TA) + \sum_{y \neq x} Pr(\theta^{(n+1)} = y, \bar{TA}|\theta^{(n)} = x) \\ &= Q(x,x) + \sum_{y \neq x} Q(x,y)[1 - \min\{1, p_y/p_x\}] \end{aligned}$$

The first step to obtaining the stationary distribution of this chain is to prove that the probabilites p_x satisfy the reversibility condition. For $x = y$, equation (4.10) is trivially satisfied. For $x \neq y$, suppose first that $p_y > p_x$. Then

$$p_x P(x,y) = p_x Q(x,y) = Q(y,x)\min\left\{1, \frac{p_x}{p_y}\right\}p_y = p_y P(y,x)$$

Analogous calculations follow for the case $p_y < p_x$. Therefore, the chain is reversible and the probabilities p_x, $x \in S$ provide the stationary distribution of the chain.

If Q is aperiodic, so will be P and the stationary distribution is also the limiting distribution. It is not difficult to find examples of symmetric transition matrices. The random walk chain (Example 4.1) with f symmetric around 0 is an example. The birth and death model with $p_x = q_{x+1}$ is another example.

4.7 Continuous state spaces

This section considers sequences of random quantities that form a Markov chain in R but still retain a discrete parameter space T. Although not explored in as many textbooks as the case of discrete state spaces, it may be found in a few recent books (Gillespie, 1992; Medhi, 1994; Meyn and Tweedie, 1993). There are a few changes required with respect to the discrete case but the main results of the previous sections are still valid. In particular, convergence to the limiting distribution, the ergodic theorem and the central limit theorem need basically technical changes in the conditions of the chain to hold.

4.7.1 Transition kernels

Markov chains are still defined in terms of Equation (4.1). If the conditional probabilities do not depend on the step n, the chain is homogeneous. Then the transition kernel $P(x, A)$ (section 4.2) is again used to define the chain. The analogy with the discrete case breaks when trying to consider $P(x, \{y\})$, which is always null in the continuous case and not useful in this context. Therefore, transition matrices cannot be constructed and transition kernels must be used instead. However, given that $P(x, \cdot)$ defines a probability distribution, the notation $P(x, y)$ can be used as

$$P(x, y) = Pr(\theta^{(n+1)} \leq y \mid \theta^{(n)} = x) = Pr(\theta^{(1)} \leq y \mid \theta^{(0)} = x), \text{ for } x, y \in S$$

when P is absolutely continuous with respect to y. Also associated with this conditional distribution, one can obtain the conditional density

$$p(x, y) = \frac{\partial P(x, y)}{\partial y}, \text{ for } x, y \in S.$$

This density can be used to define the transition kernel of the chain instead of $P(x, A)$. The state space S does not need to be the entire line. It can be any interval or collection of intervals for results below to hold.

The conditional transition probability over m steps is given by

$$P^m(x, y) = Pr(\theta^{(m+n)} \leq y \mid \theta^{(n)} = x), \text{ for } x, y \in S,$$

the transition kernel over m steps is given by

$$p^m(x, y) = \frac{\partial P^m(x, y)}{\partial y}, \text{ for } x, y \in S$$

and the equivalent equation to (4.2) has the form

$$P^{n+m}(x, y) = \int_{-\infty}^{\infty} P^m(z, y) p^n(x, z) dz, \quad m, n \geq 0.$$

This is the continuous version of the Chapman-Kolmogorov equations. For

$m = 1$, it reduces to

$$P^{n+1}(x, y) = \int_{-\infty}^{\infty} P(z, y) p^n(x, z) dz, \quad n \geq 0.$$

The marginal distribution at any step n has density $\pi^{(n)}$ ($n \geq 0$) that can be obtained from the marginal distribution at the previous step as

$$\pi^{(n)}(y) = \int_{-\infty}^{\infty} p(x, y) \pi^{(n-1)}(x) dx. \tag{4.11}$$

Example 4.1 (continued) Assume now that the random displacements w_n are continuous quantities with common density $f(w)$ and $\pi^{(0)}$ is some continuous distribution. As in the discrete case,

$$\theta^{(n+1)} = \theta^{(n)} + w_{n+1} = w_1 + \ldots + w_{n+1}.$$

Note that this is the model used for the system equation in Example 2.8 with $f(\cdot) = f_N(\cdot; 0, W)$, assuming a constant system variance W. The transition probabilities are

$$
\begin{aligned}
P(x, y) &= Pr(\theta^{(n)} + w_{n+1} \leq y \mid \theta^{(n)} = x) \\
&= Pr(w_{n+1} \leq y - x) \\
&= \int_{-\infty}^{y-x} f(w) dw
\end{aligned}
$$

and $p(x, y) = f(y - x)$. The marginal distribution at each step is recursively obtained as

$$\pi^{(n)}(y) = \int_{-\infty}^{\infty} f^n(y - x) \pi^{(0)}(x) dx$$

where f^n is the nth convolution of f. In the case of Example 2.8, these calculations simplify due to the normality assumptions to give $\pi^{(n)} = N(a, R + nW)$ if $\pi^{(0)} = N(a, R)$.

Example 4.9 A numerical sequence $y_1, y_2, \ldots$ is split by bars whenever $y_j > y_{j+1}$ and starts with a bar. A run is a collection of numbers limited by bars. For example, the portion 3,6,9,2,3,1,5,2 of a sequence is split as

$$\mid 3, 6, 9 \mid 2, 3 \mid 1, 5 \mid 2.$$

and (3,6,9), (2,3) and (1,5) are runs.

Consider a sequence $\phi_1, \phi_2, \cdots$ of independent random variables with identical distribution $U[0, 1]$ and let $(\theta^{(n)})_{n \geq 1}$ be a Markov chain on $S = (0, 1)$ formed by the initial values of the runs. The transition kernel of the chain is obtained from

$$P(x, y) = \sum_{m=1}^{\infty} Pr(\theta^{(n+1)} \leq y, \psi_n = m \mid \theta^{(n)} = x)$$

where ψ_n is the length of the nth run. It can be shown that this leads to

$$p(x,y) = \begin{cases} e^{1-x} & \text{, if } y < x \\ e^{1-x} - e^{y-x} & \text{, if } y > x \end{cases}$$

Assume now that $\theta^{(n)} = (\theta_1^{(n)}, ..., \theta_d^{(n)})'$ is a random vector in R^d. A sequence $(\theta^{(n)})_{n \geq 0}$ in $S \subset R^d$ is a Markov chain with continuous state space if

$$Pr(\theta^{(n+1)} \leq y \mid \theta^{(n)} = x, \theta^{(n-1)} = x^{(n-1)}, ..., \theta^{(0)} = x^{(0)})$$
$$= Pr(\theta^{(n+1)} \leq y \mid \theta^{(n)} = x),$$

where $x^{(0)}, ..., x^{(n-1)}, x$ and $y \in R^d$ and $z \leq w$ for d-dimensional vectors stands for $z_i \leq w_i$, $i = 1, ..., d$. Homogeneous chains are defined in the same way and transition probabilities are given by

$$P(x,y) = Pr(\theta^{(n+1)} \leq y \mid \theta^{(n)} = x) = Pr(\theta^{(1)} \leq y \mid \theta^{(0)} = x).$$

As this transition defines a d-dimensional conditional distribution, the transition kernel given by the conditional density associated with this distribution is

$$p(x,y) = \frac{\partial P(x,y)}{\partial y}.$$

4.7.2 Stationarity and limiting results

The stationary or invariant distribution π of a chain with transition kernel $p(x,y)$ must satisfy

$$\pi(y) = \int_{-\infty}^{\infty} \pi(x)\, p(x,y)\, dx \tag{4.12}$$

which is the continuous version of Equation (4.5). The interpretation remains the same, as is clear from Equation (4.11).

To study convergence and limiting results, the classification of states must be revisited. Instead of considering hitting time to a given state x, one must consider hitting time T_A to a given set $A \subset S$ and a distribution ν. A chain is said to be ν-irreducible if for a set A with positive probability under ν, $\rho_{xA} = Pr_x(T_A < \infty) > 0$, for all $x \in S$. This is equivalent to imposing the existence of an integer n such that $P^n(x, A) > 0$. A chain is irreducible if there is at least one distribution ν ensuring that it is ν-irreducible. Usually, irreducibility is simpler to verify through this last condition, with $\nu = \pi$. Also, for most of the chains of interest for simulation, $P(x, A) > 0$.

The other vital properties for establishment of limiting results are aperiodicity and positive recurrence. These are defined as in the discrete case but considering sets A with positive probability under ν to replace atoms $\{y\}$ for which transition probabilities are always null in the continuous case. For the specific case of recurrence, a slightly stronger notion of Harris recurrence is used to replace positive recurrence (Tierney, 1994). Ergodic chains are defined as aperiodic, Harris recurrent chains.

Once these definitions are given, all important convergence results established for discrete chains are valid here. For convenience, they are reviewed below for a continuous Markov chain $\theta^{(n)}$ with state space $S \subset R^d$ and stationary distribution π:

- Irreducibility and aperiodicity of the chain is equivalent to ergodicity and the uniqueness of π as the limiting distribution in total variation norm.

- The ergodic averages of real-valued functions $t(\theta)$ converge almost surely to their limiting expectations (when they exist) as stated in (4.6).

- The central limit theorem stated in (4.9) with τ^2 given by (4.8) applies to ergodic averages.

Verification of ergodicity may be difficult for some chains. Tierney (1994) proved that most Markov chains used nowadays for simulation are ergodic and the above results can be applied. Finally, the important condition of reversibility of a chain is given by

$$\pi(x)p(x,y) = \pi(y)p(y,x), \text{ for all } x, y \in S \qquad (4.13)$$

in direct analogy with the discrete case. Note that (4.12) follows directly from (4.13) by integrating both sides with respect to x. Reversible chains have proved to be very useful in helping specification of a Markov chain with limiting distribution π.

4.8 Simulation of a Markov chain

Consider an ergodic Markov chain $(\theta^{(n)})_{n \geq 0}$ with state space $S \subset R^d$, transition kernel $p(x,y)$ and initial distribution $\pi^{(0)}$. Generation of a value of this chain starts with a value for $\theta^{(0)}$ sampled from $\pi^{(0)}$. The value of $\theta^{(1)}$ is then distributed with density $p(\theta^{(0)}, \cdot)$ and can be generated from it (section 1.3). For $\theta^{(2)}$, this procedure is repeated by drawing from a distribution with density $p(\theta^{(1)}, \cdot)$. Iterating this scheme through the steps of the chain leads to drawing $\theta^{(n)}$ from a distribution with density $p(\theta^{(n-1)}, \cdot)$, for all n.

As the value of n gets large, the draws become increasingly closer to draws from the limiting distribution π and can be considered as approximate draws from π. Note that all chain values sampled after convergence is reached are also draws from π due to stationarity. Here and throughout this book, convergence is assumed to hold approximately for an iteration whose marginal distribution is arbitrarily close to the equilibrium distribution π and not in the formal sense.

These simple results have a far-reaching impact on simulation well beyond the study of Markov chains. They provide sophisticated machinery with which to approach sampling from any (possibly highly dimensional) distribution π. This area of study is collectively known as Markov chain Monte Carlo (MCMC) methods. Chapter 2 evidenced the need for summarization of posterior distributions π. Many examples showed that this task is far from trivial in complex models and Chapter 3 presented some approximating alternatives, including simulation. It was shown there that non-iterative techniques have a limited scope. As the dimension of the model gets large, they become less reliable.

This chapter provides the means by which sampling from virtually any posterior distribution π can be approached. One simply has to embed π as the limiting distribution of an ergodic Markov chain with transition kernel p. The main requirement from $p(x, \cdot)$ is to provide distributions that can be sampled from.

The remainder of this book is devoted to presenting and studying Markov chains whose simulation lead to draws from a limiting distribution of interest π. Many important questions arise and will be tackled during the next chapters. The most basic one is whether such chains can always be constructed and sampled from. It is remarkable that there are many such chains for any posterior distribution π, no matter how complex the model is. Examples 4.7 and 4.8 have provided simple cases that hint at a positive answer. Although these examples considered only the discrete context, they will be extended in the next chapters to accommodate highly dimensional continuous distributions.

Other relevant questions regard the criteria for selecting the iteration to stop sampling and choice of kernel among the possible alternatives. The first point deals with determination of convergence of the chain. In more precise terms, one would wish to ascertain how close the marginal distribution $\pi^{(n)}$ of the current iteration is to the target distribution π. Many aspects are involved including theoretical bounds on probability distances and the computational complexity of calculations required. In broad terms, the initial chain values are far from the stationary distribution and should be discarded. This period is referred to as the warm-up or burn-in period for obvious reasons. Deletion of these values hopefully improves the accuracy of ergodic averages but enlarges their variance by the reduction of the sample size. This point will be returned to when convergence diagnostics are discussed in the next chapter. In any case, Markov chain simulation opens up a host of possibilities for sampling in situations where the direct sampling methods of Chapter 1 and the indirect sampling methods of Chapter 3 do not apply.

Before concluding this section, it is important to comment on methods for assessing the accuracy of the estimates provided by ergodic averages as measured by their variance τ_n^2/n or its limiting approximation τ^2/n. There are many methods reviewed by Ripley (1987) and Geyer (1992). They can be broadly divided into direct, time series and batching methods. Assume that interest lies in the estimation of $E_\pi[t(\theta)]$ and a stream of simulated values $t^{(n)} = t(\theta^{(n)})$ is available from a Markov chain with stationary distribution π.

Direct methods are based on estimates $\hat\tau_1^2$ of τ^2 obtained by respective replacements of σ^2 and ρ_k in (4.8) by moment estimates $\hat\sigma^2 = \hat\gamma_0$ and $\hat\rho_k = \hat\gamma_k/\hat\gamma_0$, $k \leq k^*$ where

$$\hat\gamma_k = \frac{1}{n} \sum_{j=1}^{n-k} t^{(j)} t^{(j+k)} - \bar{t}^2 \text{ , for } k \geq 0$$

and for $k > k^*$, $\hat\rho_k = 0$ where k^* is chosen to limit the sum to include relevant terms. Unfortunately, the resulting estimate $\hat\tau^2$ will not be consistent. Many variants of the above estimates considering other multiplying constants instead of $1/n$ have been proposed and are reviewed by Priestley (1981). Consistency is ensured by appropriately downweighting higher order autocorrelation (Geyer, 1992).

There are many methods of estimation based on time-series ideas. One approach is to fit an autoregressive structure to the time series $t^{(n)}$ and estimate τ^2 from the estimated residual variance. More generally, ARMA models can be used. Geweke (1992) used estimates based on the spectral density $S(w)$ of the series evaluated at frequency $w = 0$. These are commonly used in the study of time series in the frequency domain (Priestley, 1981). Geyer (1992) considers other estimators based on the autocovariance structure of Markov chains.

Batching estimators are based on the simple idea of dividing the stream of $n = mk$ chain values into k batches of m successive values. The rationale behind it is to seek approximate independence between batches and therefore take the batch averages $\bar{t}_1, ..., \bar{t}_k$ as approximately independent quantities. The value of k is chosen to enforce approximate independence or at least very low autocorrelation of the sequence. Generally, values of k should be in the region 10-30 (Schmeiser, 1982). Then, for large m, each $\bar{t}_i$ has common approximate mean $E_\pi[t(\theta)]$ and variance $\tau^2/m = kVar(\bar{t})$. Hence, the sample variance of the $\bar{t}_i$ estimates $kVar(\bar{t})$ and τ^2 is estimated as

$$\frac{\hat\tau_2^2}{n} = \frac{1}{k(k-1)} \sum_{i=1}^{k} (\bar{t}_i - \bar{\bar{t}})^2$$

where $\bar{\bar{t}} = \bar{t}$. Inference about $E_\pi[t(\theta)]$ is based on an approximate sampling distribution $\sqrt{n}\{\bar{t} - E_\pi[t(\theta)]\}/\hat\tau_2 \sim t_{k-1}(0,1)$, to account for the extra variability due to the estimation of τ^2. The simplicity of the method has made it a popular choice for estimation of sampling variance in Markov chains.

4.9 Data augmentation or substitution sampling

This chapter concludes with an example of a Markov chain constructed by Tanner and Wong (1987) to have π as a limiting distribution. Assume that $\theta = (\phi, \psi)$ has posterior distribution π and components ϕ and ψ can have any dimension. Assume also that interest lies mostly in inference about ϕ, ψ being a set of constructed parameters, latent variables or additional data. Under the last interpretation data already available is augmented by ψ, hence the first name of the method.

The marginal posterior densities of ϕ and ψ are obtained as

$$\pi(\phi) = \int \pi(\phi|\psi)\pi(\psi)d\psi$$

$$\pi(\psi) = \int \pi(\psi|x)\pi(x)dx$$

where x is a dummy argument playing the role of the parameter ϕ. Substituting the second into the first equation and interchanging integration signs leads to

$$\pi(\phi) = \int \pi(\phi|\psi)\left[\int \pi(\psi|x)\pi(x)dx\right]d\psi$$

$$= \int p(x, \phi)\pi(x)dx \qquad (4.14)$$

where $p(x, \phi) = \int \pi(\phi|\psi)\pi(\psi|x)d\psi$. Equation (4.14) suggests that successive substitutions of $\pi(\phi)$ form an iterative algorithm. Since the integrations involved will generally not be feasible analytically, they can be replaced by sampling approximations. Hence the second name of the method. More importantly, Equation (4.14) is Equation (4.12), satisfied by the stationary distribution $\pi(\phi)$ of a Markov chain $\phi^{(n)}$ with transition kernel $p(x, \phi)$.

The iterative solution proposed by Tanner and Wong (1987) is to update an approximation $\pi^{(n)}$ of $\pi(\phi)$ to $\pi^{(n+1)}$ as follows:

1. Draw a sample $\phi_1, ..., \phi_m$ from $\pi^{(n)}(\phi)$.
2. Draw a sample $\psi_1, ..., \psi_m$ from $\pi(\psi)$. This is approximately achieved by drawing ψ_i from $\pi(\psi|\phi_i)$, $i = 1, ..., m$ (section 1.3).
3. Form a Monte Carlo approximation

$$\pi^{(n+1)}(\phi) = \frac{1}{m}\sum_{i=1}^{m}\pi(\phi|\psi_i)$$

For large m, these steps form an sampling-based approximation to an iteration of the transition kernel $p(x, \phi)$, that only depends on the conditional distributions. It has been stressed in the previous chapters that despite the difficulty in direct sampling from the marginal distributions, sampling from the conditionals is generally easy for many models. So, step

2 can in these cases be performed. Step 3 informs that sampling required at step 1 will be from an approximation to the marginal given by a discrete mixture of conditionals. If it is easy to sample from the conditionals, then all sampling procedures are easily carried out.

Tanner and Wong (1987) showed that provided the transition kernel $p(x, \phi) > 0$ for all pairs of points (x, ϕ) in the support of $\pi(\phi)$ and is uniformly bounded and equicontinuous, the data augmentation algorithm is uniformly ergodic and $\pi(\phi)$ is the unique distribution satisfying (4.14). These results are valid for any choice of m. Also, note that the algorithm is symmetric in terms of ϕ and ψ. Therefore, after convergence, samples of size m are available from the marginal distributions of both ϕ and ψ.

It is also interesting to consider the case $m = 1$. An iteration of the algorithm simply alternates single draws from the conditional distributions. This structure resembles Example 4.7 and is the basis of Gibbs sampling scheme, introduced in the next chapter. Also, this iterative algorithm can be extended to more than two components (Gelfand and Smith, 1990).

4.10 Exercises

1. Obtain the transition matrices for the chains described in Examples 4.1 to 4.4.

2. Consider the problem of sending binary messages of length d through a channel consisting of various stages. The transmission through each stage has error probability α. Let θ_0 be the message originally sent and θ_n the message received at the nth stage.

   (a) Obtain the transition matrix.

   (b) What is the probability that a signal is correctly received at the second stage?

   (c) What is the probability that a signal is incorrectly received for the first time at the second stage?

3. Let P be a transition matrix. Prove that P^k is stochastic and has at least one eigenvalue equal to one, $k \geq 1$.

4. Consider the Ehrenfest model for $r = 3$.

   (a) Obtain $Pr_x(T_0 = n)$ for $x \in S$ and $1 \leq n \leq 3$.

   (b) Obtain the matrices P, P^2 and P^3.

   (c) If $\pi_0 = (1, 1, 1, 1)/4$, calculate π_1, π_2 and π_3.

5. Consider a modified Ehrenfest model with $S = \{0, 1, \cdots, r\}$ and transi-

tion probabilities given by

$$
P(x, y) = \begin{cases}
(d - x)/2d & \text{, if } y = x + 1 \\
1/2 & \text{, if } y = x \\
x/2d & \text{, if } y = x - 1 \\
0 & \text{, if } |y - x| \neq 1
\end{cases}
$$

If initially the first urn has on average $d/2$ balls, how many balls can be expected to lie in the first urn at the next step?

6. Show that

$$
E(T_y \mid \theta^{(0)} = x) = \sum_{n=0}^{\infty} Pr_x(T_y > n) \quad \text{and} \quad E(N(y) \mid \theta^{(0)} = x) = \sum_{n=1}^{\infty} P^n(x, y)
$$

7. Show that if $y \in S$ is a transient state then, for all $x \in S$,

$$
Pr_x(N(y) < \infty) = 1 \quad \text{and} \quad E[N(y) \mid \theta^{(0)} = x] = \frac{\rho_{xy}}{1 - \rho_{yy}} < \infty
$$

Show also that if $y \in S$ is a recurrent state then

$$
Pr_y(N(y) = \infty) = 1 \quad \text{and} \quad E[N(y) \mid X_0 = y] = \infty
$$

8. Show that birth and death processes are irreducible Markov chains when $p_x > 0$ for $x \geq 0$ and $q_x > 0$ for $x > 0$.

9. Consider again a birth and death process in $S = \{0, 1, 2, \cdots\}$ with $p_x = \frac{x+2}{2(x+1)}$ and $q_x = \frac{x}{2(x+1)}$.

   (a) Determine whether the chain is recurrent or transient.
   (b) Determine $Pr_x(T_a < T_b)$ for $a < x < b$.

10. Show that

    (a) $\rho_{xy} > 0$ is equivalent to $P^n(x, y) > 0$ for some $n > 0$;
    (b) if $x \in S$ is recurrent, $x \to y$ and $y \to x$ then $y \in S$ is also recurrent;
    (c) if $x \in S$ is recurrent, $x \not\to y$ and $y \not\to x$ then $P(x, y) = 0$.

11. Consider the 2×2 version of the Gibbs sampler presented in Example 4.7. Obtain the 4×4 transition matrix P and show that π is the stationary distribution of the chain.

12. The exports of a country can be modelled, under economic stability, as a Markov chain with states $+1$, 0 and -1 representing, respectively, growth of 5% or more, variation smaller than 5% and decline of 5% or more with respect to the previous year. Let the transition matrix be

$$
P = \begin{pmatrix}
0.8 & 0.2 & 0 \\
0.35 & 0.3 & 0.35 \\
0 & 0.4 & 0.6
\end{pmatrix}
$$

(a) Does this chain have a limiting distribution?

(b) Determine the average return times for all states.

13. Points 0,1,2,3 and 4 are marked clockwise in a circle. At each step, a particle moves with probability p to the right (clockwise) and $1 - p$ to the left (anti-clockwise). Let $\theta^{(n)}$ be the position of the particle in the circle at the nth step. Obtain the transition matrix and the limiting distribution, if it exists. What is the expected number of steps for a return to the initial state?

14. Analyse the limiting behaviour of the transition matrices

$$P_1 = \begin{pmatrix} 1 & 0 \\ 0 & 1 \end{pmatrix} \; , \quad P_2 = \begin{pmatrix} 0 & 1 \\ 1 & 0 \end{pmatrix} \quad \text{and} \; P_3 = \begin{pmatrix} \frac{1}{2} & \frac{1}{2} \\ 0 & 1 \end{pmatrix}$$

15. Prove the ergodic theorem for $t(\theta) = \theta_i$, $i = 1, 2$, under the conditions of Example 4.7 with $p_{00} = p_{11} = p/2$ and $p_{01} = p_{10} = (1 - p)/2$. In other words, obtain that

$$Pr\left(\lim_{j \to \infty} \frac{1}{j} \sum_{l=1}^{j} \theta_{il} = \frac{1}{2} \right) = 1 \; , \quad i = 1, 2$$

16. Consider a Markov chain $\theta^{(n)}$ and define the autocovariance of lag k $(k \geq 0)$ of the chain as $\gamma_k = Cov_\pi(\theta^{(n)}, \theta^{(n+k)})$, the variance of $\theta^{(n)}$ as $\sigma^2 = \gamma_0$, the autocorrelation of lag k as $\rho_k = \gamma_k/\sigma^2$ and $\tau_n^2/n = Var_\pi(\bar\theta_n)$. Show that

$$\tau_n^2 = \sigma^2 \left(1 + 2 \sum_{k=1}^{n-1} \frac{n-k}{n} \rho_k \right)$$

and that $\tau_n^2 \to \tau^2$ as $n \to \infty$ where $\tau^2 = \sigma^2(1 + 2\sum_{k=1}^{\infty} \rho_k)$ if the series of autocorrelation is summable.

17. Obtain for continuous state spaces that

(a) $P^{n+m}(x, y) = \int_{-\infty}^{\infty} P^m(z, y) p^n(x, z) dz$;

(b) $\pi^{(n)}(y) = \int_{-\infty}^{\infty} p(x, y) \pi^{(n-1)}(x) dx = \int_{-\infty}^{\infty} p^n(x, y) \pi^{(0)}(x) dx$.

18. Consider a chain $(\theta^{(n)})_{n \geq 1}$ formed according to the process described in Example 4.9. Show that the transition kernel and the density of the limiting distribution of the chain are respectively given by

$$p(x, y) = \begin{cases} e^{1-x} & , \text{ if } y < x \\ e^{1-x} - e^{y-x} & , \text{ if } y > x \end{cases} \quad \text{and } \pi(y) = \begin{cases} 2(1-y) & , \text{ if } 0 < y < 1 \\ 0 & , \text{ otherwise} \end{cases}.$$

19. Discuss the sense in which an iteration of the data augmentation method provides a sampling-based version of the transition kernel in (4.14).

## CHAPTER 5

Gibbs sampling

5.1 Introduction

This chapter introduces the first widely used class of schemes for stochastic simulation using Markov chains. It is generically known as Gibbs sampling because it originated in the context of image processing. In this context, the posterior of interest for sampling is a Gibbs distribution. Borrowing concepts from Mechanical Statistics, the density of the Gibbs distribution can be written as

$$f(x_1, ..., x_d) \propto \exp\left[-\frac{1}{kT}E(x_1, ..., x_d)\right] \qquad (5.1)$$

where k is a positive constant, T is the temperature of the system, E is the energy of the system, a positive function, and x_i is the characteristic of interest for the ith component of the system, $i = 1, ...d$. In Mechanical Statistics, x_i is the position or perhaps the velocity and position of the ith particle and in image processing, it is (an indicator of) the colour of the ith pixel of an image.

The energy function E is commonly given by a sum of potential functions V. These sums operate over collections of subgroups of components over which each potential function is evaluated. The subgroups generally obey some neighbouring relationship in their definition. This leads to a probability specification based on local properties, useful for modelling spatial interaction between components. The main drawback is the difficulty in determination of the global properties, such as the normalizing constant.

Geman and Geman (1984) discuss this modelling problem extensively with special regard for sampling schemes and comparison with Markov random fields*. Their sampling scheme explored the conditional structure implied by the local specification. Even though it was a well known and influential paper in the area, their paper was not published in a mainstream statistical journal. This is one of the few possible explanations for the delay in the introduction of their powerful results for the solution of Bayesian problems in general. Gelfand and Smith (1990) were the first authors to successfully point out to the statistical community at large that the sampling scheme devised by Geman and Geman (1984) for Gibbs dis-

---

* Markov random fields are extensions of Markov chains obtained when the parameter set T is multidimensional.

tributions could in fact be used for a host of other posterior distributions. In that sense, it is somewhat misleading that the scheme retained the name Gibbs sampling and Robert (1994) proposed to change it to Bayesian sampling. The paper by Gelfand and Smith (1990) also compared the Gibbs sampling scheme with the data augmentation algorithm (section 4.9) and sampling-importance resampling (section 3.5).

This chapter tries to describe the development of the area up to now. Some questions are still not entirely settled and there is a risk of obsolescence involved. The Gibbs sampling algorithm is described in the next section and some of its main properties presented. Section 5.3 deals with the description of implementation and convergence acceleration techniques. As previously discussed, one of the main difficulties when sampling via Markov chains is the verification of convergence of the chain. This problem is addressed in section 5.4 where statistical techniques for convergence monitoring and identification are introduced. Most of the material from these two sections can be applied to any MCMC scheme, not just the Gibbs sampler. Section 5.5 applies Gibbs sampling to hierarchical and dynamic models. Finally, the chapter provides a brief description of the main software available for Bayesian inference using Gibbs sampling.

5.2 Definition and properties

Gibbs sampling is a MCMC scheme where the transition kernel is formed by the full conditional distributions. Assume as before that the distribution of interest is $\pi(\theta)$ where $\theta = (\theta_1, ..., \theta_d)'$. Each one of the components θ_i can be a scalar, a vector or a matrix[†]. Consider also that the full conditional distributions $\pi_i(\theta_i) = \pi(\theta_i | \theta_{-i})$, $i = 1, ..., d$ are available. This means that they are completely known and can be sampled from.

The problem to be solved is to draw from π when direct generation schemes are costly, complicated or simply unavailable but when generations from the π_i are possible. Gibbs sampling provides an alternative generation scheme based on successive generations from the full conditional distributions. It can be described in the following way:

1. Initialize the iteration counter of the chain $j = 1$ and set initial values $\theta^{(0)} = (\theta_1^{(0)}, ..., \theta_d^{(0)})'$.

2. Obtain a new value $\theta^{(j)} = (\theta_1^{(j)}, ..., \theta_d^{(j)})'$ from $\theta^{(j-1)}$ through successive generation of values

$$\theta_1^{(j)} \sim \pi(\theta_1 | \theta_2^{(j-1)}, ..., \theta_d^{(j-1)})$$
$$\theta_2^{(j)} \sim \pi(\theta_2 | \theta_1^{(j)}, \theta_3^{(j-1)}, ..., \theta_d^{(j-1)})$$

[†] The reader may prefer to think about them as scalars if that helps. This point is readdressed in section 5.3.4.

$$\vdots$$

$$\theta_d^{(j)} \sim \pi(\theta_d | \theta_1^{(j)}, ..., \theta_{d-1}^{(j)})$$

3. Change counter j to $j + 1$ and return to step 2 until convergence is reached.

When convergence is reached, the resulting value $\theta^{(j)}$ is a draw from π. As the number of iterations increases, the chain approaches its equilibrium condition. Convergence is then assumed to hold approximately.

This presentation where each iteration consists of a single change to all components is favoured by Gelfand and Smith (1990). The original work of Geman and Geman (1984) presented a chain with iterations formed by a change to a given component. Step 2 is obtained in the special case where the components are changed in a fixed and constant order.

Example 5.1 (Carlin, Gelfand and Smith, 1992) Let $y_1, ..., y_n$ be a sample from a Poisson distribution for which there is a suspicion of a change point m along the observation process where the means change, $m = 1, ..., n$. Given m, the observation distributions are $y_i | \lambda \sim Poi(\lambda)$, $i = 1, ..., m$ and $y_i | \phi \sim Poi(\phi)$, $i = m + 1, ..., n$. The model is completed with independent prior distributions $\lambda \sim G(\alpha, \beta)$, $\phi \sim G(\gamma, \delta)$ and $m \sim uniform$ over $\{1, ..., n\}$ where α, β, γ and δ are known constants. The posterior density is

$$
\begin{aligned}
\pi(\lambda, \phi, m) \quad &\propto \quad f(y_1, ..., y_n | \lambda, \phi, m)\, p(\lambda, \phi, m) \\
&= \quad \prod_{i=1}^{m} f_P(y_i; \lambda) \prod_{i=m+1}^{n} f_P(y_i; \phi) \quad f_G(\lambda; \alpha, \beta) f_G(\phi; \gamma, \delta) \frac{1}{n} \\
&\propto \quad \prod_{i=1}^{m} e^{-\lambda} \lambda^{y_i} \prod_{i=m+1}^{n} e^{-\phi} \phi^{y_i} \lambda^{\alpha-1} e^{-\beta\lambda} \phi^{\gamma-1} e^{-\delta\phi} \\
&\propto \quad \lambda^{\alpha + \sum_{i=1}^{m} y_i - 1} e^{-(\beta+m)\lambda} \phi^{\gamma + \sum_{i=m+1}^{n} y_i - 1} e^{-(\delta+n-m)\phi}
\end{aligned}
$$

from where it becomes simple to obtain the full conditional densities

$$
\pi_\lambda(\lambda) = G\left(\alpha + \sum_{i=1}^{m} y_i, \beta + m\right), \quad \pi_\phi(\phi) = G\left(\gamma + \sum_{i=m+1}^{n} y_i, \delta + n - m\right) \quad \text{and}
$$

$$
\pi_m(m) = \frac{\lambda^{\alpha + \sum_{i=1}^{m} y_i - 1} e^{-(\beta+m)\lambda} \phi^{\gamma + \sum_{i=m+1}^{n} y_i - 1} e^{-(\delta+n-m)\phi}}{\sum_{l=1}^{n} \lambda^{\alpha + \sum_{i=1}^{l} y_i - 1} e^{-(\beta+l)\lambda} \phi^{\gamma + \sum_{i=l+1}^{n} y_i - 1} e^{-(\delta+n-l)\phi}}, \quad m = 1, ..., n
$$

All these distributions are easily sampled from (see Chapter 1) and the iterative scheme repeating steps 1-3 can be operated without difficulty.

The obvious form to obtain a sample of size n from π is to replicate n chains until convergence. Alternatively, after convergence all draws from a chain come from the stationary distribution. Therefore n successive values from this chain after the burn-in period will also provide a sample from π. The issue of how to form a sample is readdressed in more detail in the next section.

A typical trajectory of a Gibbs sampling chain is illustrated in Figure 5.1. An iteration is completed after d moves along the coordinate axes of the components of θ. Convergence diagnostics are complex as d can be very large and will be left for section 5.4. Note that the convergence must be in distribution which means that the joint distribution of all parameter components must converge to the joint posterior for all values of θ. This exhaustive verification is far from trivial.

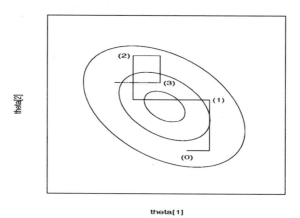

theta[1]

Figure 5.1. *Typical trajectory of the Gibbs sampler in a bidimensional parametric space $d = 2$. The concentric curves represent the contour lines of the posterior density. The points in parentheses represent a possible trajectory of the chain.*

A few basic facts must be established beforehand. First of all, Gibbs sampling defines a Markov chain. This is clearly the case as the probabilistic change at iteration j depends only on chain values at step $j - 1$. Also, the chain is homogeneous as the transitions are only affected by the iteration through the chain values. It is not difficult to obtain the transition kernel as

$$p(\theta, \phi) = \prod_{i=1}^{d} \pi(\phi_i | \phi_1, ..., \phi_{i-1}, \theta_{i+1}, ..., \theta_d) \qquad (5.2)$$

which clearly depends on the iterations only through the chain values θ and ϕ. This chain with a complete scan over all components is not reversible although each individual change is reversible. Peter Green pointed out that the chain can be made reversible by taking each iteration of the chain to consist of the complete scan through the components followed by another scan through the components in reversed order (Besag, 1986).

Another important result is the derivation that the equilibrium distribution of a chain with transition kernel (5.2) is π. This result was derived in a special case of $d = 2$ discrete components in Example 4.7. In the continuous case, the same argument cannot be used but the mechanics of the algorithm is the same. If a Markov chain with transition kernel $p(\theta, \phi)$ has limiting distribution π^∞, then the stationarity condition (4.12) must be satisfied by p and π^∞ and the chain must be irreducible. Irreducibility is easy to verify for each application by checking that $P(x, A) > 0$ for all sets A with positive posterior probability. For statistical applications, it is generally satisfied but see Example 5.2 below for a counterexample.

To check stationarity, let $\theta = (\theta_1, \theta_2)$ with marginal limiting densities $\pi^\infty(\theta_1)$ and $\pi^\infty(\theta_2)$. The limiting full conditional distribution of θ_1 is $\pi_1^\infty(\theta_1)$. The transition kernel (5.2) simplifies to

$$p(\theta, \phi) = \pi(\phi_1|\theta_2)\pi(\phi_2|\phi_1)$$

where $\phi = (\phi_1, \phi_2)$ has components with the same dimensions as those of θ. As $\int \int \pi(\phi_2|\phi_1)\pi^\infty(\theta_1|\theta_2)d\theta_1 d\phi_2 = \int \pi(\phi_2|\phi_1)d\phi_2 \times \int \pi^\infty(\theta_1|\theta_2)d\theta_1 = 1$,

$$
\begin{aligned}
\pi^\infty(\theta_2) &= \int \int \pi(\phi_2|\phi_1)\pi^\infty(\theta_1|\theta_2)\pi^\infty(\theta_2)d\theta_1 d\phi_2 \\
&= \int \int \pi(\phi_2|\phi_1)\pi^\infty(\theta)d\theta_1 d\phi_2 \quad (5.3)
\end{aligned}
$$

Integrating (4.12) with respect to ϕ_2 gives the marginal limiting density of ϕ_1 as

$$
\begin{aligned}
\pi^\infty(\phi_1) &= \int \int p(\theta, \phi)\pi^\infty(\theta)d\theta d\phi_2 \\
&= \int \int \int \pi(\phi_1|\theta_2)\,\pi(\phi_2|\phi_1)\pi^\infty(\theta)d\theta_1 d\theta_2 d\phi_2 \\
&= \int \pi(\phi_1|\theta_2)\pi^\infty(\theta_2)d\theta_2 \quad (5.4)
\end{aligned}
$$

where the last equality follws from (5.3). The only distribution satisfying (5.4) must have $\pi(\phi_1|\phi_2) = \pi^\infty(\phi_1|\phi_2)$. The same argument could be used to give $\pi(\phi_2|\phi_1) = \pi^\infty(\phi_2|\phi_1)$ and the limiting distribution must have the same full conditionals as the posterior. The same argument follows for θ divided into d blocks of components as these can always be rearranged in two blocks θ_i and θ_{-i}. This means that all limiting full conditionals are given by the posterior full conditionals. This does not in general guarantee that $\pi^\infty = \pi$ (see Exercise 5.2 for an example where it fails). Nevertheless, Besag (1974) showed that under very mild conditons, the set of all full conditional distributions determine the joint distribution. Therefore, the Markov chain with transition kernel (5.2) converges to the distribution of interest π and the iterative sampling scheme with steps 1-3 above draws in the limit a value from this distribution.

Formal convergence conditions for the Gibbs sampler were established by Roberts and Smith (1994) and Tierney (1994). The results are presented in terms of continuous parameter spaces but can be extended for combinations of continuous and discrete parameters (Example 5.1). A simple example of a reducible chain where convergence fails is given below.

Example 5.2 (O'Hagan, 1994; Roberts, 1996) Consider $\theta = (\theta_1, \theta_2)$ uniformly distributed over two disjoint regions $A = A_1 \times A_2$ and $B = B_1 \times B_2$ of the plane with probabilities p_A and p_B adding up to 1. Consider also that A_1 and B_1 are disjoint regions on the θ_1 axis and A_2 and B_2 are disjoint regions on the θ_2 axis. This implies that full conditionals are also uniform but over regions that depend on the starting point. Chains that start in A_1 will lead to sampling θ_2 uniformly over A_2 which implies sampling θ_1 uniformly over A_1 and this situation perpetuates itself. Points from B will never be reached. Analogously, chains starting in B_1 will never reach points in A. This chain is clearly reducible and will have uniform limiting distribution over A or B depending on the starting point.

5.3 Implementation and optimization

Despite the theoretical results ensuring the convergence of the Gibbs sampler, its practical implementation may be complicated by the potential complexity of the models considered. Convergence of the sampler becomes difficult to characterize. Given that it is a numeric and iterative method, practical strategies to improve the efficiency of the method may have a considerable impact on its computational cost. Efficiency broadly consists of reducing the number of burn-in iterations and the amount of arithmetic operations required at each iteration. The techniques presented are related to the basic MCMC methods as described in the previous chapter and represented by the Gibbs sampler. More general techniques using other forms of chains will be presented in Chapter 7.

5.3.1 Forming the sample

The previous section presented two forms to obtain a sample of size n from the posterior distribution π. The obvious one is to process n chains in parallel until convergence, say after m iterations, and take as sample elements the mth chain value from each of the n chains. The generation procedure will then require mn generations from the chain. If chains are initialized independently, the sample consists of independent values from π. Independence is easier to establish if the initial values are all different and preferably with larger dispersion than in the posterior (section 5.4.3).

Another form is to consider a single chain and explore ergodic results. After convergence, all chain values have marginal distribution given by the equilibrium distribution π. So, a sample of size n may be formed by n sucessive values from this chain. This generation will require $m + n$ generations from the chain. This is substantially less than independent sampling. The difficulty here is that the sample elements are no longer independent due to chain dependence. Ergodic theorems ensure that inference based on this sample is still valid. From a practical point of view, there may be problems if the chain autocorrelation is too high and the sample is not large enough to acknowledge it. In these cases, chains may take too long to adequately cover the entire parameter space appropriately. As a result, some relevant regions may be underrepresented in the sample.

An alternative approach accommodating independence is to take for the sample chain values at every kth iteration after the burn-in period. Markovian processes only have first order dependence. As the lag between iterations increases, chain values become less and less correlated and are virtually independent for a large enough value of the lag k. A sample of size n with *quasi*-independent elements thus requires $m + kn$ generations from the chain. The value of k is typically smaller than m and again an improvement over independent sampling is obtained. There is no gain in efficiency, however, by this approach and estimation is shown below to be always less precise than retaining all chain values. This procedure is advantageous if computer storage of values is limited. Useful indicators of dependence are given by the chain autocorrelations (section 4.8). A formal approach for selection of k is given in section 5.4.4.

Another compromise is to take a small number l, say less than 10, of independent chains, run them until convergence and then retain from each of them n/l successive values from the sample. This will lead to a sample of size n obtained after $l[m + (n/l)] = lm + n$ generations from the chains. Yet another variant is obtained by retaining every kth chain value after convergence with a total of $l[m + (n/l)k] = lm + kn$ generations. In computational terms, there are efficiency losses with respect to using a single chain and gains with respect to independent sampling.

The independent sampling approach was suggested by Gelfand and Smith (1990) and used by some authors shortly afterwards. The single chain approach was emphatically advocated by Geyer (1992) backed by ergodic theorems. Sampling every kth iteration was discussed by Raftery and Lewis (1992). Gelman and Rubin (1992a) recommended the use of a small number of independent chains backed by an example from Gelman and Rubin (1992b) where single chains provide indication of convergence of ergodic averages to different limits. Gelman et al. (1995) argue that somehow the benefits from *quasi*-independent sampling are diluted when running few chains.

There is no general agreement on the subject although it is generally agreed that running n parallel chains in practice is computationally inefficient and unnecessary. The main debate is whether a few parallel chains are needed. If the convergence properties of the chain are well understood then clearly a single chain suffices. As these characteristics are hard to obtain, prudence suggests that a few pilot parallel chains should be run. If they quickly settle around common values then a single chain can be safely used to extract a large sample for inference. Otherwise, there may be minor characteristics of the posterior distribution such as secondary modes far from the mode that require very large samples to be noticed. In this case, these parallel chains should be run longer and their values should be retained for the sample. Convergence diagnostics are the subject of the next section and these points will be returned to in more detail there.

5.3.2 Scanning strategies

The Gibbs sampler described in the previous section involved a complete scan over the components. All iterations consisted of visits to update the components in the same deterministic order, typically $1 \to 2 \to \cdots \to d$. There are many other possible scanning or updating strategies for visiting the components of θ.

Geman and Geman (1984) proved convergence to the joint distribution in a discrete setting for all visiting schemes that guarantee that all components are visited i.o. when the chain is run indefinitely. The reversible Gibbs sampler where at each iteration each component is visited in a fixed order and then visited again in reversed order satisfies this property. In this case, each iteration consist of $2d$ updates and comparisons between strategies should bear that in mind.

Another scheme where an i.o. schedule is guaranteed draws a number i from $\{1, ..., d\}$ with fixed positive probabilities at each iteration and only updates the θ_i at that iteration. To make it more comparable with the determinsitic scan, an iteration of these random scans can be defined by a collection of d such updates.

Roberts and Sahu (1997) consider a random permutation scan where at each iteration a permutation of $\{1, ..., d\}$ is selected and components are visited in that order. Zeger and Karim (1991) describe a Gibbs sampling scheme where some components were visited only every kth iteration. This also guarantees an i.o. visiting schedule for fixed, finite k.

Assume now that π is a multivariate normal distribution with precision matrix $\Phi = (\phi_{ij})$. For this setting, Roberts and Sahu (1997) showed that convergence for the deterministic scan is faster than for the random scan if Φ is tridiagonal $(\pi(\theta_i|\theta_{-i}) = \pi(\theta_i|\theta_{i-1}, \theta_{i+1})$, for all $i)$ or if Φ has nonnegative partial correlations $(\phi_{ij} \leq 0)$. This result is particularly important because both dynamic and hierarchical models lead to tridiagonal matrices if variances are known. Their results also indicate that more precise distributions lead to faster convergence both for the deterministic and random scans.

5.3.3 Using the sample

Whatever the scheme chosen for forming the sample, after it is used a sample of vectors $\theta_1, ..., \theta_n$ generated from the posterior distribution π is available. Assume also the more general case where these are successive values from a single Markov chain. A sample from the ith component of θ is given by $\theta_{1i}, ..., \theta_{ni}$. Marginal point or interval summaries of any real function $\psi = t(\theta)$ are estimated by their corresponding estimators based on the sample. This is always a consistent estimator by the ergodic theorem (4.6). The quality of this estimator can be judged by the central limit theorem (4.9) from where approximate confidence intervals about the MCMC estimates may be formed.

So, the posterior mean of ψ is estimated by $\hat{E}(\psi) = \hat{\psi} = (1/n)\sum_{j=1}^{n}\psi_j$ where $\psi_j = t(\theta_j)$, $j = 1, ..., n$. The posterior variance of ψ is similarly estimated by noting that $\sigma_\psi^2 = Var(\psi) = E(\psi^2) - [E(\psi)]^2$. Each expectation is estimated by an application of (4.6) and σ_ψ^2 is estimated by $\hat{\sigma}_\psi^2$ where

$$\hat{\sigma}_\psi^2 = \hat{E}(\psi^2) - [\hat{E}(\psi)]^2 = \frac{1}{n}\sum_{j=1}^{n}(\psi_j - \hat{\psi})^2,$$

the sample variance. The denominator n may be replaced by $n-1$ but this change is irrelevant for two reasons. First, typically n is large which makes the change immaterial. Second, it does not remove the bias of the estimator as usually happens when independent sampling is performed.

Consider again the problem of choosing between a sample of n successive values and a sample of $m = n/k$ values obtained by skipping every kth iteration. Note that k such sub-samples with *quasi*-independent draws are formed. Denote by $\hat{\psi}_1, ..., \hat{\psi}_k$ the averages of the sub-samples and $\hat{\psi} = (1/k)\sum_{j=1}^{k}\hat{\psi}_j$ the average over the complete sample. There are $k+1$ estimators of $E(\psi)$ and they are all consistent estimators by the ergodic theorem. It can be shown that $Var(\hat{\psi}) \leq Var(\hat{\psi}_j)$, for all j (O'Hagan, 1994; MacEachern and Berliner, 1994). This means that independence sampling comes at the expense of reduced efficiency.

Credibility intervals are similarly obtained by estimating the interval limits by the respective sample quantiles. As in section 3.5, if $n = 1000$, and an equal tails 95% probability interval for ψ is required, it can be estimated by the interval with limits given by the 25th and 975th largest sample values of ψ. Again, these values are consistent estimators of the 0.025 and 0.975 quantiles of ψ by (4.6).

All above estimators have a sampling distribution that is approximated by (4.9). The asymptotic variance of these estimators, which is different from the posterior variance, can be estimated by the methods described in section 4.8. The central limit theorem ensures that estimation errors are $O(n^{-1/2})$.

The marginal densities $\pi(\theta_i)$ can be estimated by (a smoothed version of) the histogram of sampled values of θ_i (section 3.5). Better estimators can be obtained by using conditional distributions. Recalling that $\pi(\theta_i) = \int \pi(\theta_i|\theta_{-i})\pi(\theta_{-i})d\theta_{-i}$, a Monte Carlo estimator is given by

$$\hat{\pi}(\theta_i) = \frac{1}{n}\sum_{j=1}^{n}\pi(\theta_i|\theta_{j,-i}) \qquad (5.5)$$

where the $\theta_{j,-i}$, $j = 1, ..., n$ are a sample from the marginal $\pi(\theta_{-i})$. Again, the ergodic theorem ensures that $\hat{\pi}$ is a consistent estimator and it obeys a central limit theorem for every value of θ_i. These estimators are always continuous for continuous parameters. More importantly, they are based on information about the form of the posterior. For that reason, Gelfand and Smith (1990) call it a Rao-Blackwellized density estimator. This is a reference to the Rao-Blackwell theorem that states that estimators are always improved (in the sense of reducing sampling variance) by conditioning on sufficient statistics. They proved the result for density estimation in the context of independent sampling. The general proof of the result for Markov chain sampling is given by Liu, Wong and Kong (1994). The same idea can be used to obtain better estimates of moments of $t(\theta_i)$ through

$$\hat{E}[t(\theta_i)] = \frac{1}{n}\sum_{j=1}^{n}E[t(\theta_i)|\theta_{j,-i}]$$

although the gains are not as large here.

5.3.4 Reparametrization

Going back to Figure 5.1, an iteration is formed by moves along the coordinate axes of the components of θ. If there is weak dependence between the components, the moves will be ample. The chain will then move freely through the parametric space and convergence will be fast. An extreme case is posterior independence between the components. The full conditionals are equal to the marginals and convergence is immediate.

Often, the posterior structure leads to high correlation between some of the components of θ (section 5.5.2). Figure 5.2 illustrates this point for a bidimensional parameter. The contours of the posterior show strong dependence between the components of θ and chain moves, governed by the conditional densities, will be small. The chain will take many iterations to adequately cover the parametric space and as a result convergence is slow. In this case, the Gibbs sampler will be inefficient. Examples can be constructed in larger dimension models where convergence can be slowed to any arbitrary amount of iterations (Shephard, 1994).

A simple and sometimes effective way to reduce convergence time is to use reparametrizations. This point was discussed in Chapter 3 in the context of improving approximations. Adequate transformations in the parameter space may produce situations of near independence that are ideal for fast convergence of the chain. Unfortunately, there are no rules to determine suitable transformations but frequently linear transformations that produce a diagonal variance matrix provide good results. Two important classes of models where these transformations can be found are presented below.

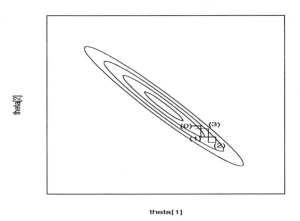

theta[1]

Figure 5.2. *Contour lines of a bivariate posterior density with components highly correlated. A possible chain trajectory is also depicted to illustrate slow convergence.*

Example 5.3 For the regression model described in section 2.3, the conditional posterior was $\pi(\beta|\phi) = N(b_\phi, B_\phi)$ and therefore posterior correlations depend on the posterior variance B_ϕ. Numerical techniques described in section 1.4 can be applied to obtain the square root matrix A_ϕ such that $A_\phi A'_\phi = B_\phi^{-1}$. So, $\alpha = A_\phi \beta \sim N(A_\phi b_\phi, I_d)$ given ϕ and the components of vector α are independent a posteriori given ϕ.

In the case of simple linear regression $y_i = \beta_1 + \beta_2 x_i + e_i$ with non-informative prior $p(\beta) \propto k$, α has components $\alpha_1 = \beta_1 + \beta_2 \bar{x}$ and $\alpha_2 = \beta_2$. This is equivalent to centering covariates and working with the model $y_i = \alpha_1 + \alpha_2(x_i - \bar{x}) + e_i$. Depending on how close are the values of the x_i, the plot of the posterior conditional density of $\beta|\phi$ will present contour lines that are as concentrated as those exhibited in Figure 5.2. In multiple linear regression, centering covariates is usually enough.

The conditional dependence between components of α is removed and the only remaining posterior correlation is between α and ϕ. Sampling components of α is as simple as sampling components of β but the chain will converge faster. Once α is sampled, a sample value of β is obtained by $\beta = A_\phi^{-1}\alpha$.

Example 5.4 Consider the hierarchical (or random effects) model

$$y_{ij} = \mu + \alpha_i + e_{ij} \quad , \quad e_{ij} \sim N(0, \sigma^2)$$

where the random effects α_i associated with the observation groups have distribution $\alpha_i \sim N(0, \tau^2)$, $j = 1, ..., n_i$, $i = 1, ..., m$. Note that by writing $\beta_i = \mu + \alpha_i$ and $\beta = \mu$, the one-way classification model of Example 2.7 is recovered. Assume also a non-informative prior $p(\mu) \propto k$. It can be shown that given variance components σ^2 and τ^2, posterior correlations between model parameters are given by

$$Cor(\mu, \alpha_i) = -\left(1 + \frac{\sigma^2/n_i}{\tau^2/m}\right)^{-1/2} \quad \text{and} \quad Cor(\alpha_i, \alpha_j) = \left(1 + \frac{\sigma^2/n_i}{\tau^2/m}\right)^{-1}$$

High posterior correlations occur if $\sigma^2/n_i \ll \tau^2/m$. Roberts and Sahu (1997) showed that asymptotically (as $m \to \infty$) the deterministic scan over the parameters has faster convergence than random scans.

Gelfand, Sahu and Carlin (1995) suggest the hierarchical parametrization with the β_i replacing the α_i and show that given the variance components σ^2 and τ^2, posterior correlations between model parameters are given by

$$Cor(\mu, \beta_i) = -\left(1 + \frac{m\tau^2}{\sigma^2/n_i}\right)^{-1/2} \quad \text{and} \quad Cor(\beta_i, \beta_j) = \left(1 + \frac{m\tau^2}{\sigma^2/n_i}\right)^{-1}$$

Now, low correlations are obtained for the conditions of high correlations in the original parametrization. So, depending on the data structure, it is more appropriate to work on another parametrization. Theoretical support is provided by Roberts and Sahu (1997). Gelfand, Sahu and Carlin (1995) argue that in practice the presence of random effects implies excess randomness and therefore it is expected that σ^2/n_i will be smaller than τ^2/m, which would justify the use of the β_i. They extended their approach to nested random effects models. Generalized linear models with random effects were studied by Gelfand, Sahu and Carlin (1996) and, even though analytic results are no longer available, they arrived at the same qualitative recommendations.

Vines, Gilks and Wild (1994) suggest the reparametrization $\nu = \mu - \bar{\alpha}$ and $\xi_i = \alpha_i - \bar{\alpha}$ and show that given the variance components σ^2 and τ^2, posterior correlations between model parameters are given by

$$Cor(\nu, \xi_i) = 0 \quad \text{and} \quad Cor(\xi_i, \xi_j) = -\frac{1}{m}$$

Now, correlations have the advantage of not depending on the variance components. Again, the idea can be extended to more general models. For this parametrization, convergence with the random permutation scans is faster than for all other scanning strategies.

These examples suggest a general strategy based on approximate posterior normality (section 3.2). The approximate variance V is a first order approximation for the posterior variance. Its square root matrix A can be calculated and a linear transformation $\alpha = A^{-1}\theta$ operated. This will provide approximate posterior independence to the first order. In more general models, in addition to the computational cost of finding A, there is also the added cost of sampling α instead of θ. Other approximations to the posterior variance may be sought. Hills and Smith (1992) suggest using the sample variance obtained from a pilot chain.

Another simple but important point is to observe the structure of the model and parameters. For example, these transformations will provide a better result if the posterior for each parameter behaves like the normal distribution. Variance parameters will not have this behaviour unless a large number of observations is collected. Otherwise, a logarithmic transformation is recommended before application of the orthogonalization procedures above. Optimal strategies for some parameters should not be blindly applied for other parameters.

5.3.5 Blocking

So far, nothing has been said about the choice of components that form the parameter vector θ. In principle, the way the components are arranged in blocks of parameters is completely arbitrary and includes as a special case blocks formed by scalar components. The structure of Gibbs sampling, also illustrated in Figure 5.1, makes moves according to the coordinate axes of the blocks. Scalar blocks lead to moves along each component of θ. Larger blocks allow moves in more general directions. This can be very beneficial computationally when there is high correlation between components. The slow, componentwise moves may be replaced by fast moves incorporating the information about dependence between components. These moves are dictated by the joint full conditional for the block of parameters considered, which incorporates the correlation structure. This intuitive consideration is confirmed by the theoretic results of Liu, Wong and Kong (1994). They showed that estimates obtained by blocking components are generally more precise than those obtained by treating each component separately.

Derivations of Roberts and Sahu (1997) show that, for random scans, convergence improves as the number of blocks decreases. Thus, blocking is beneficial. They also proved that blocking is beneficial for non-negative partial correlation distributions and more so as the partial correlation of the components in the block gets larger. These results were obtained only for a multivariate normal π and extrapolations should be made with caution. They also provided an example where blocking worsens convergence.

Although it is hard to determine optimal blocking strategies, some basic rules should be followed. When a parametric vector or matrix is specified in block, they generally have joint full conditionals that are easy to sample from. The important message is to block as much as possible for sampling. Of course, if the complete parameter vector forming a single block could be sampled, there would be no need for Gibbs sampling! Therefore, the only restriction is the ability to sample from the full conditional distributions formed.

5.3.6 Sampling from the full conditional distributions

In some cases, the form of the full conditional distribution is not recognizable which prevents sampling via the conventional algorithms. Chapter 1 presented a host of other general-purpose options such as (adaptive) rejection and reweighted sampling methods. According to Carlin and Louis (1996), these situations are an indication that an altogether different approach should be applied instead of insisting on Gibbs sampling.

Ritter and Tanner (1992) developed yet another sampling scheme from difficult full conditionals. Their approach is similar to adaptive rejection by being based on the evaluation of the full conditional at a few selected points. For that reason, they called it the griddy Gibbs sampler. Let $\pi_i(\theta_i)$ be a difficult full conditional distribution. Then, sampling from π_i can be approximately performed as follows:

1. Take a grid of points $\theta_{i1}, ..., \theta_{im}$, evaluate $\pi_i(\theta_{ij})$, $j = 1, ..., m$, and normalize them to obtain weights $w_1, ..., w_m$.

2. Use the weights $w_1, ..., w_m$ to construct a simple approximation to the distribution function of π_i.

3. Draw a value from π_i by the probability integral transform method (section 1.3).

There are many possibilities for the construction in step 2. The simplest one is to use piecewise constant functions (discrete distribution). Piecewise linear functions are also easy to sample from and allow for continuous sampling. Higher order polynomials and even splines may be used but it is important to keep it simple to sample from. Tanner (1993) suggests that the number of points m should be kept small for the burn-in period of the chain and doubled after convergence for good approximations only when it matters. He also suggests an adaptive scheme to revise the selected grid to include more points in higher density regions.

5.4 Convergence diagnostics

As previously discussed, a value from the distribution of interest π is only obtained when the number of iterations of the chain approaches infinity. In practice this is not attainable and a value obtained at a sufficiently large iteration is taken instead of being drawn from π. The difficulty is the determination of how large this iteration should be. There is no simple answer to this question and most efforts have been directed at studying as close as possible the convergence characteristics of the chain. Most results below can be applied to any MCMC method although for a few of them the use of Gibbs sampling is required.

There are two main ways to approach the study of convergence. The first one is more theoretical and tries to measure distances and establish bounds on distribution functions generated from a chain. In particular, one can study the total variation distance between the distribution of the chain at iteration j and the limiting distribution π. Special aspects derived from the probabilistic structure of the chain can also be studied. This approach was pursued by Meyn and Tweedie (1994), Polson (1996), Roberts and Polson (1994), Roberts and Tweedie (1994) and Rosenthal (1993) to cite just a few papers (see also the references in those papers). This is an area that is certainly going to grow as we increase our understanding of the subject. At the moment, however, the results have had little impact on practical work (Cowles and Carlin, 1996).

The study of convergence of the chain can also be approached from a statistical perspective, i.e. by analysing the properties of the observed output from the chain. This is an empirical as opposed to a theoretical treatment of the problem and is obviously more practical. The difficulty with this approach is that it can never guarantee convergence because it is only based on observations from the chain (see Example 5.5 below).

Although the two approaches to the study of convergence are valid and complement each other, theoretical results have proved to be more difficult to obtain and apply to practical problems. This book will provide a more detailed description of the convergence diagnostics based on the statistical properties of the observed chain. Cowles and Carlin (1996) and Brooks and Roberts (1995) provide comparative and illustrative reviews of many of these methods. Robert (1995) reviews some possibilities involving the two approaches.

5.4.1 Rate of convergence

Geman and Geman (1984) showed for the discrete case that the Gibbs sampler is a uniformly ergodic Markov chain. Uniform ergodicity determines an exponential rate of convergence of the chain to the limiting distribution and could be taken as an indication of fast convergence. However, there is no indication of control over the rate of convergence and the Gibbs sampler can have an extremely slow convergence in some cases. The example below illustrates this point in a very simple context.

Example 5.5 (O'Hagan, 1994) Consider again the situation of Example 4.7 where $\theta = (\theta_1, \theta_2)'$ is bivariate and $\pi(\theta)$ is given by the table of probabilities below

		θ_2	
		0	1
θ_1	0	$p/2$	$(1-p)/2$
	1	$(1-p)/2$	$p/2$

Observe that $\pi(\theta_i) = bern(1/2)$, $i = 1, 2$, and the posterior correlation between θ_1 and θ_2 is $\rho = 2p - 1$.

Using properties of the Gibbs sampler, it is easy to obtain that $Pr(\theta_1^{(j)} = 1|\theta_1^{(j-1)} = 1) = Pr(\theta_1^{(j)} = 0|\theta_1^{(j-1)} = 0) = p^2 + (1-p)^2$ and, consequently, $Pr(\theta_1^{(j)} = 1|\theta_1^{(j-1)} = 0) = Pr(\theta_1^{(j)} = 0|\theta_1^{(j-1)} = 1) = 2p(1-p)$. Taking $p_j = Pr(\theta_1^{(j)} = 1)$ gives $p_j = \rho^2 p_{j-1} + b$ where $b = 2p(1-p)$. The solution for p_j is $p_j = \rho^{2(j+1)} p_0 + b(1 - \rho^{2(j+1)}))/(1 - \rho^2)$.

The transition matrix formed by the marginal chain $(\theta_1^{(j)})_{j \geq 0}$ has eigenvalues 1 and ρ and therefore the rate of convergence is $|\rho|$. Ergodicity of the chain is ensured if $p > 0$ but if p is close to 1 or 0, this rate will be close to 1, the chain will tend not to move and convergence to the limiting distribution is very slow. In the limit, $p_j \to b/(1 - \rho^2) = 1/2$ as expected. However, if $p = 0.999$, $\rho = 0.998$ and, after 100 iterations, $p_{100} = 0.667p_0 + 0.165$ which is still far from the appropriate limit.

The point raised in this example is far from rare in many applications and although correlations as high as 0.998 are not common, a similar effect is obtained with high dimensional parameter spaces with much smaller correlations. Once again, it seems sensible in these cases to reparametrize the model. This point was already discussed in the previous sections and will be returned to in section 5.5.

5.4.2 Informal convergence monitors

Gelfand and Smith (1990) suggested a few informal checks of convergence based on graphical techniques. After m iterations in n parallel chains, a histogram of the n values of the mth iterates of a given function of θ can be plotted. This function can be one of the components of θ and the histogram may be smoothed if desired. The procedure is repeated after a further k iterates are obtained in the chains. The value of k need not be large if one suspects convergence after m iterations. It cannot be low as the chain correlation will still be affecting possible similarities of the histograms. Typically, values between 10 and 50 are reasonable. Convergence is accepted if the histograms cannot be distinguished.

Same ideas can be used with a single chain. A trajectory of the chain exhibiting the same qualitative behaviour through iterations after a transient initial period is an indication of convergence. Similarly, the trajectory of the ergodic averages can be evaluated and plotted. An asymptotic behaviour over many successive iterations indicates convergence. Figure 5.3 shows the ergodic averages of variance components in a nested random effects model (Gamerman, 1997). The indication of convergence for both components seems to be very clear.

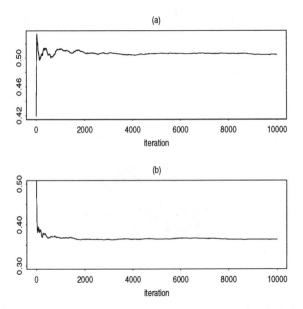

Figure 5.3. *Ergodic averages of two parameters with number of iterations of the chain. The parameters are standard deviations of random effects at: (a) - individual; (b) - unit level in a longitudinal study of epilepsy treatment (Gamerman, 1997).*

Similar ideas can be used with graphical representations of the simulated values of a few chosen (transformations of) parameters. The resulting plots provide a rough indication of stationarity behaviour when the sequence of values tends to concentrate around the same pattern. This visual impression can be reinforced when chains started at different values oscillate in the same region.

Example 5.6 Souza (1997) considers a number of hierarchical and dynamic models to describe the nutritional pattern of pregnant women. The data depicted in Figure 5.4 consist of the weight gains of 68 pregnant women at 5 to 7 visits to the Instituto de Puericultura e Pediatria Martagão Gesteira from the Universidade Federal do Rio de Janeiro. One of the simplest models she adopted was the simple hierarchical regression on time where

$$y_{ij}|\alpha_i, \beta_i, \phi \sim N(\alpha_i + \beta_i t_{ij}, \sigma^2) \quad , \quad j = 1, ..., n_i, \ i = 1, ..., 68$$
$$\alpha_i|\alpha \sim N(\alpha, \tau_\alpha^{-1}) \quad , \quad i = 1, ..., 68$$
$$\beta_i|\beta \sim N(\beta, \tau_\beta^{-1}) \quad , \quad i = 1, ..., 68$$
$$(\alpha, \beta)' \sim N((0, 0)', 10^3 I_2),$$

prior independent scale parameters σ^{-2}, τ_α and $\tau_\beta \sim G(0.001, 0.001)$ and y_{ij} and t_{ij} are the jth weight measurement and visit time of the ith women, $j = 1, ..., n_i, \ i = 1, ..., 68$.

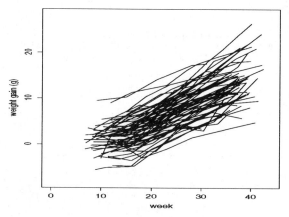

Figure 5.4. *Data on the weight gain of pregnant women (Souza, 1997).*

The Gibbs sampler can be applied after calculation of the full conditional distributions. These are all normal and gamma distributions which can be easily sampled from. Figure 5.5 shows traces of some model parameters for two parallel chains started at different points. It seems to indicate convergence after around 1500 iterations.

These techniques must be used with caution and should always be accompanied by some theoretical reasoning. Graphical techniques may be deceptive indicating constancy that may not be so evident under a different scale. More importantly, there are many chains that exhibit every indication of convergence without actually achieving it. They are called metastable chains and are the subject of much research in probability theory (Capocaccia, Cassandro and Olivieri, 1977; Cassandro et al., 1984).

Example 5.5 (continued) It was established that if p is close to 1 or 0, the chain will tend not to move. Therefore, the observed trajectory will have long constant stretches indicating a metastable behaviour. This is known not to be an indication of convergence because $\pi(\theta_i = j) = 1/2$, $i = 1, 2$, $j = 0, 1$, and convergence will only be achieved when the chain begins to alternate values 0 and 1 with very similar frequencies.

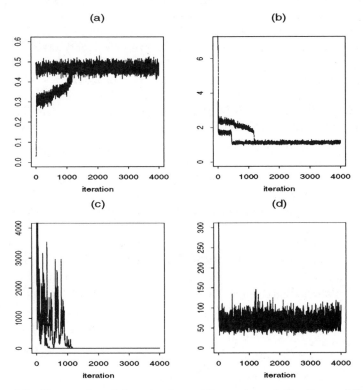

Figure 5.5. *Traces of model parameters with number of iterations of the two chains. The parameters are: (a) - β, the populational growth; (b) - σ, the observational standard deviation; (c) - τ_α, the precision of the population of intercepts; (d) - τ_β, the precision of the population of regression coefficients.*

5.4.3 Convergence prescription

Raftery and Lewis (1992) proposed a method to establish the length of a chain required for a MCMC run. More specifically, the methodology suggests values of m, the number of burn-in iterations, k, the number of iterations to be skipped between stored chain values and n, the size of the sample values that must be stored to achieve a given Monte Carlo precision of estimates.

The setting for these choices is the estimation of u, the q quantile of a given function $\psi = t(\theta)$, i.e. $q = Pr_\pi(\psi \le u)$. The method requires that the Monte Carlo estimate $\hat{q}$ satisfies $Pr(|\hat{q} - q| \le r) = s$. A common choice is the tail probability with $q = 0.025$ in which case u is the lower limit of the equal tail 95% posterior credibility interval for ψ. One may require that the value of this probability be estimated in a MCMC run with error smaller than $r = 0.01$ with confidence $s = 0.99$. So, 95% posterior intervals would be given by intervals with posterior probabilities between 93% and 97% with 99% confidence. This confidence level is due to the estimation of q by MCMC and should not be confused with posterior uncertainty about ψ, governed by π.

This problem is tackled at the simpler level of a binary chain $Z^{(j)} = I(\psi^{(j)} \le u)$ where $\psi^{(j)}$ is the value of ψ at the jth iteration of the MCMC for θ. $Z^{(j)}$ is derived from a Markov chain but is not a Markov chain. It is reasonable to assume however that dependencies between iterations fall off quickly with lag and new chains $Z_k^{(j)}$ may be formed by taking the values of $Z^{(j)}$ at every kth iteration. The value of k is chosen as the smallest lag to make the first order Markov chain preferable to the second order Markov chain for the chain $Z_k^{(j)}$. A test using the BIC (Schwarz, 1978) is used to choose between the two models for every value of k (Raftery and Lewis, 1992).

Once the value of k is chosen, the next step is to determine the number $m = m^*k$ iterations to be discarded. This is done by choosing m^* such that at iteration m^*, the chain $Z_k^{(j)}$ has marginal distribution arbitrarily close to the limiting distribution implied by π. This is equivalent to requiring that $|Pr(Z_k^{(m^*)} = 1|Z_k^{(0)} = j) - Pr_\pi(\psi \le u)| < \epsilon$ and $|Pr(Z_k^{(m^*)} = 0|Z_k^{(0)} = j) - Pr_\pi(\psi > u)| < \epsilon$, $j = 0, 1$. This is obtained using the results from Example 4.5 as

$$m^* = \frac{\log\left(\frac{\epsilon(\alpha+\beta)}{\max\{\alpha,\beta\}}\right)}{\log|1 - \alpha - \beta|}$$

where α and β are the $(0, 1)$ and $(1, 0)$ elements of P_k and P_k is the transition matrix of the chain $Z_k^{(j)}$ (Exercise 5.8).

The value of $n = n^*k$ is chosen using a central limit theorem for the chain $Z_k^{(j)}$. Note that q is estimated by the ergodic average $\hat{q} = \bar{Z}_{k,n} = (1/n)\sum_{j=1}^n Z_k^{(m+j)}$, that has asymptotic variance given by $\tau^2 = \alpha\beta(2 - \alpha - \beta)/(\alpha + \beta)^3$. The approximating distribution for $\hat{q}$ gives

$$n^* = \left(\frac{\tau z_{(1+s)/2}}{r}\right)^2$$

where z_γ is the γ quantile of the $N(0,1)$ distribution. As an assessment of the magnitudes involved, in the most favourable case of independent sampling with $k = 1$, estimation of the quantile 0.025 with largest error 0.0125 with 95% confidence level requires $n = 600$ iterations. When the error is reduced to 0.005, the number of iterations required increases to 3746.

So, one must specify the quantile of interest q, the convergence tolerance ϵ, the estimation tolerance r and confidence level s in advance. In addition, a pilot run must be observed to estimate the values of α and β. An appropriate run length will then be prescribed. This run can be used to refine the estimates of α and β, suggesting an iterative procedure.

This diagnostic obviously depends on the chosen ψ and quantile. Raftery and Lewis (1996) recommend using it for all quantities of interest with $q = 0.025$ and $q = 0.975$ as the tails are harder to estimate in general. These provide a collection of values of k, m and n and the largest of each is chosen. If l chains are to be used, then each should be run for $m + n/l$ iterations to ensure convergence. Note that the procedure does not require any information about the chain itself, just its output. Brooks and Roberts (1995) provide an example of a slow convergence chain where severe under- and over-estimation of the required length are observed. They point out the use of marginal indicators and estimation of α and β as the weak points of the method and suggest its use alongside other convergence diagnostics.

Table 5.1. Convergence prescription summary for data on pregnant women

Parameter	k	m	n
β	1(3)*	2(18)	3866 (15759)
σ	1 (1)	3 (2)	4112 (3946)
τ_α	1 (1)	3 (3)	4285 (4112)
τ_β	2 (1)	8 (2)	8128 (3787)

* - values in parentheses refer to the second chain.

Example 5.6 (continued) Two chains were run for 4000 iterations to obtain the prescribed values of the spacing k, the burn-in period m and the sample size n retained for inference. Based on the observed chain values these are given for the two chains according to values in Table 5.1 with $q = 0.025$, $r = 0.005$ and $s = 0.95$. They seem to indicate that the chain lengths used are generally appropriate. Figure 5.5 suggests that the burn-in periods have been mostly underestimated. Note also that different starting values may provide large variation in the prescribed values.

5.4.4 Formal convergence methods

As in the previous subsection, the methods presented here diagnose convergence based on exploration of the statistical properties of the observed chain. The methods here attempt to decide whether convergence can be safely assumed to hold rather that prescribing the run length to achieve convergence. There have been many methods presented in the literature. Most of them are covered by the review papers of Brooks and Roberts (1995) and Cowles and Carlin (1996). Only a few of the most cited and used in the literature are presented here.

Time series analysis

Consider a real function $\psi = t(\theta)$ and its trajectory $\psi^{(1)}, \psi^{(2)}, \dots$ obtained from $\psi^{(j)} = t(\theta^{(j)})$, $j = 1, 2, \dots$ This trajectory defines a time series and ergodic averages of this series can be evaluated. Geweke (1992) suggests the use of tests on ergodic averages to verify convergence of the chain based on the series $\psi^{(j)}$.

Assume observation of the chain for $m + n$ iterations and form averages

$$\bar{\psi}_b = \frac{1}{n_b} \sum_{j=m+1}^{m+n_b} \psi^{(j)} \text{ and } \bar{\psi}_a = \frac{1}{n_a} \sum_{j=m+n-n_a+1}^{m+n} \psi^{(j)}$$

where $n_b + n_a < n$. If m is the length of the burn-in period, then $\bar{\psi}_a$ and $\bar{\psi}_b$ are the ergodic averages at the end and beginning of the convergence period and should behave similarly. As n gets large and the ratios n_a/n and n_b/n remain fixed then

$$z_G = \frac{\bar{\psi}_a - \bar{\psi}_b}{\sqrt{\hat{V}ar(\psi_a) + \hat{V}ar(\psi_b)}} \to N(0, 1) \text{ in distribution}$$

So, the standardized difference z_G between the ergodic averages at the beginning and at the end of the convergence period should not be large if convergence has been achieved. Large differences indicate lack of convergence but small differences do not imply convergence. Geweke (1992) suggests the use of values $n_b = 0.1n$ and $n_a = 0.5n$ and uses spectral density estimators for the variances. This is a univariate technique but can be applied to posterior density by taking $t(\theta) = -2 \log \pi(\theta)$. As with the Raftery and Lewis (1992) diagnostic, it requires only the output from the chain and can be used with any MCMC scheme.

Example 5.6 (continued) The values of the z_G statistic for the two chains and the four parameters are all larger than 10, confirming lack of convergence. When the first 1500 iterations are removed from both chains, the values of z_G are as given by Table 5.2. They are mostly indicating convergence. Further insight into these values is provided in Figure 5.6, displaying the values of z_G for each of the four parameters after removal of a given number of iterations. The figure seems to indicate convergence after around 1200 iterations, confirming results from Figure 5.5.

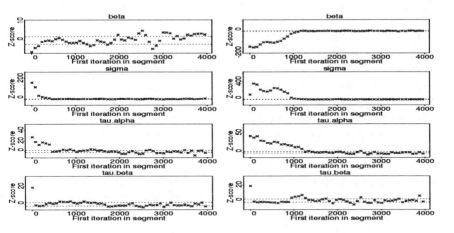

Figure 5.6. *Geweke convergence diagnostic z_G for the two chains and the four chosen parameters for the data on pregnant women plotted against the number of discarded iterations.*

Further exploration of the time series structure of the chain to study convergence of the chain has been the subject of research of a number of authors in the area of Operational Research (Heidelberger and Welch, 1983; Schruben, Singh and Tierney, 1983; and references therein). These techniques investigate similarities between the observed series and their expected behaviour under stationarity.

Table 5.2. Geweke diagnostic z_G summary for data on pregnant women

Parameter	1st chain	2nd chain
β	0.177	-1.200
σ	0.622	-2.750
τ_α	1.220	-0.789
τ_β	0.611	-0.572

Multiple chains

Another simple method to check convergence is to use parallel chains started at different points. This technique explores the same ideas used in iterative optimization to avoid convergence to local maxima. Use of multiple chains would then prevent chains getting trapped in regions around local modes. Also, slow convergence may give rise to metastable behaviour of the chains and this can be easily detected through parallel chains. Examples of this behaviour are provided by Gelman and Rubin (1992b) and Gelman (1996). After convergence, all chains will have the same quantitative and qualitative behaviour.

Gelman and Rubin (1992a) elaborated on the idea that the chain trajectories should be the same after convergence using analysis of variance techniques. The overall idea is to test whether dispersion within chains is larger than dispersion between chains. This is equivalent to the histogram of all chains being similar to all the histograms of individual chains.

The procedure starts by initializing the chains at points that are overdispersed with respect to the posterior distribution. The number of chains does not need to be too large to avoid computational waste and is typically given by single digit numbers. For components of θ restricted to an interval, two chains initialized close to the limits of the interval is an adequate choice. For continuous components, a search for the mode(s) and respective curvature(s) (section 3.2) can be set and initial states of the chains drawn from (mixtures of) Student's t distribution(s) with moment(s) matching mode(s) and curvature(s). When there is indication of posterior multimodality, it is advisable to start at least one chain from each mode. Gelman (1996) points out that it is easy to adapt programs for calculation of Gibbs samplers to maximization of the posterior. All that is required is the substitution of random moves by deterministic moves in the direction of higher posterior density.

Considering m parallel chains and a real function $\psi = t(\theta)$, there are m trajectories $\{\psi_i^{(1)}, \psi_i^{(2)}, ..., \psi_i^{(n)}\}$, $i = 1, ..., m$, for ψ. The variances between chains B and within chains W are given by

$$B = \frac{n}{m-1} \sum_{i=1}^{m} (\bar{\psi}_i - \bar{\psi})^2 \text{ and } W = \frac{1}{m(n-1)} \sum_{i=1}^{m} \sum_{j=1}^{n} (\psi_i^{(j)} - \bar{\psi}_i)^2$$

where $\bar{\psi}_i$ is the average of observations of chain i, $i = 1, ..., m$, and $\bar{\psi}$ is the average of these averages. Under convergence, all these mn values are drawn from the posterior and σ_ψ^2, the variance of ψ, can be consistently estimated by W, B and the weighted average $\hat{\sigma}_\psi^2 = (1 - 1/n)W + (1/n)B$.

If, however, the chains have not yet converged, then initial values will still be influencing the trajectories. Due to their overdispersion, they will force $\hat{\sigma}_\psi^2$ to overestimate σ_ψ^2 until stationarity is reached. On the other hand, before convergence, W will tend to underestimate σ_ψ^2 because each chain will not have adequately traversed the complete state space. Following this reasoning, an indicator of convergence can be formed by the estimator of potential scale reduction given by $\hat{R} = \sqrt{\hat{\sigma}_\psi^2 / W}$, that is always larger than 1. As $n \to \infty$, both estimators converge to σ_ψ^2 by the ergodic theorem and $\hat{R} \to 1$. Convergence can be evaluated by the proximity of $\hat{R}$ to 1. Gelman (1996) suggests accepting convergence when the value of $\hat{R}$ is below 1.2. The original estimator proposed by Gelman and Rubin (1992a) is far more elaborate and its derivation is left as an exercise. It seems that the elaboration brings unnecessary complication as there are no formal tests applied to the statistic $\hat{R}$.

Example 5.6 (continued) The values of $\hat{R}$ were evaluated for the four parameters previously chosen based on the two chains with 4000 iterations. For the scale parameters, a logarithmic transformation was used to improve the normality pattern of the posterior sample. They all lie below 1.05 providing further indication of convergence. At this stage, one can safely assume that after 2000 iterations all draws arise from the posterior distribution.

The potential scale reduction should be evaluated for all quantities of interest to provide reasonable information about convergence of the chain. Note that this is a univariate technique but, again, can be applied to the complete posterior density by taking $t(\theta) = -2 \log \pi(\theta)$.

A problem of this method is the dependence on normal theory present in the choice of initial states of the chains and formulation of variance estimators. Alternatively, non-parametric estimators of variance can be used. Also, reparametrizations may be applied to components expected to have non-normal behaviour but this increases the complexity of the verification. Another problem is the inefficiency associated with multiple chains (section 5.3.1) which should lead to very parsimonious choices of the number of chains.

Methods based on conditional distributions

Assume that θ can be divided in two blocks θ_1 and θ_2. Then, $\pi(\theta) = \pi(\theta_1|\theta_2)\pi(\theta_2) = \pi(\theta_2|\theta_1)\pi(\theta_1)$, for all θ. In the applications where Gibbs sampling can be used, full conditionals are easy to obtain but the marginal distributions are not. However, they can be estimated by (5.5) so let $\hat{\pi}(\theta_i)$ be the estimate of $\pi(\theta_i)$, $i = 1, 2$.

Zellner and Min (1995) propose two criteria for verification of convergence of the Gibbs sampler. The difference criterium is based on the statistic

$$\hat{\eta} = \pi(\theta_1|\theta_2)\hat{\pi}(\theta_2) - \pi(\theta_2|\theta_1)\hat{\pi}(\theta_1)$$

If the chain has converged, then $\hat{\eta}$ will be close to $\eta = 0$ for all θ. The ratio criterium is based on the statistics

$$\hat{\xi}_1 = \frac{\pi(\theta_2|\theta_1)\hat{\pi}(\theta_1)}{\pi(\theta_2^*|\theta_1^*)\hat{\pi}(\theta_1^*)} \text{ and } \hat{\xi}_2 = \frac{\pi(\theta_1|\theta_2)\hat{\pi}(\theta_2)}{\pi(\theta_1^*|\theta_2^*)\hat{\pi}(\theta_2^*)}$$

where $\theta^* = (\theta_1^*, \theta_2^*)'$ is another value from the state space. Both $\hat{\xi}_1$ and $\hat{\xi}_2$ are estimates of $\xi = \pi(\theta)/\pi(\theta^*)$. If the chain has converged, then $\hat{\xi}_1$ and $\hat{\xi}_2$ will be close. In addition, if they are close to ξ then the chain has converged to the correct equilibrium distribution. Zellner and Min (1995) formalize their approach by assuming a normal sampling distributions for the estimates based on (3.7). They proceed with a Bayesian analysis by assuming a vague prior distribution for the estimand, evaluating the criteria at a sample of θ values, constructing credibility intervals and testing the hypotheses of interest.

Ritter and Tanner (1992) also propose to assess convergence of the chain by looking at ratio statistics such as $\hat{\xi}_1$ and $\hat{\xi}_2$. They suggest evaluating the ratios at the chain values $\theta^{(n)}$ and plotting the histograms of the ratios. As $n \to \infty$, these histograms should become closer to a degenerate distribution at the value of 1. See also Gelfand (1992) for further discussion of analysis of histograms and Roberts (1992) for expressions of moments of ratio statistics for reversible Gibbs samplers. Again, metastable behaviour may be a problem and use of multiple chains should remedy the situation. Another problem with these methods is the need for the expression of the full conditionals which restrict their application to Gibbs samplers. It is also not clear how to split the parameter into two blocks. When the parameter dimension d is large, there are too many ways of splitting them for it to be feasible to perform convergence checks in all of them.

Other methods

There are many other methods of convergence diagnostics proposed in the literature: Liu, Liu and Rubin (1992) proposed a method based on control variables, Garren and Smith (1993) estimate the rate of convergence of chains formed by indicator variables and Johnson (1996) used coupled chains with overdispersed starting points.

Some of the advantages and disadvantages of the approaches have been discussed. The main points to consider are ease of implementation, applicability to MCMC schemes, interpretability, dependence on chain structure and availability of software (section 5.6). As previously mentioned, none of the schemes can guarantee convergence. So it is advisable that as many as possible are used in any given problem.

5.5 Applications

Applications of Gibbs sampling have been restricted so far to simple models or separate derivations of full conditional distributions. This section will provide a more complete treatment for a few special models. Inference via Gibbs sampling will be detailed for hierarchical models (section 2.4) and dynamic models (section 2.5).

5.5.1 Hierarchical models

Consider initially the 2-stage normal hierarchical model described at the beginning of section 2.4 with

$$
\begin{aligned}
y|\beta_1, \phi &\sim N(X_1\beta_1, \phi^{-1}I_n) \\
\beta_1|\beta_2, C &\sim N(X_2\beta_2, C^{-1}) \\
\beta_2 &\sim N(b, B)
\end{aligned}
$$

$$
\phi \sim G\left(\frac{n_0}{2}, \frac{n_0 S_0}{2}\right) \quad \text{independent of } C \sim W\left(\frac{n_W}{2}, \frac{n_W S_W}{2}\right)
$$

where n_0, n_W and S_0 are positive constants, b is an r-dimensional vector of constants and B and S_W are $r \times r$ and $d \times d$ positive definite matrices of constants. The parameters of the model are the d- and r-dimensional vectors β_1 and β_2 respectively, the scalar ϕ and the dispersion matrix C. Typically $r \leq d$ although this is not mathematically necessary.

The model includes as special cases the one-way classification model (Example 2.7), the random effects model (Example 5.4) and the exchangeable regression model

$$
\begin{aligned}
y_i|\beta_i, \phi &\sim N(X_i\beta_i, \phi^{-1}I_{n_i}) \quad, \quad i = 1, ..., m \\
\beta_i|\beta_2, C &\sim N(\beta_2, C^{-1}) \quad, \quad i = 1, ..., m \\
\beta_2 &\sim N(b, B)
\end{aligned}
$$

$$
\phi \sim G\left(\frac{n_0}{2}, \frac{n_0 S_0}{2}\right) \quad \text{independent of } C \sim W\left(\frac{n_W}{2}, \frac{n_W S_W}{2}\right)
$$

where $y_i = (y_{i1}, ..., y_{in_i})'$, $i = 1, ..., m$. The analysis for this model using Gibbs sampling is described and illustrated in Gelfand, Hills, Racine-Poon and Smith (1990).

Other versions of this model are possible having, for example, an unknown observational dispersion matrix or the dispersion matrix C premultiplied by ϕ. Also, the prior distributions can be changed to other nonconjugate forms or more stages can be included.

The full conditional distributions for the blocks β_1, β_2 and ϕ were obtained in section 2.4 as:

1. $\beta_1|\beta_2, \phi, C \sim N(b_\phi, B_\phi)$;

2. $\beta_2|\beta_1, \phi, C \sim N(\mu^*, C_2^*)$;

3. $\phi|\beta_1, \beta_2, C \sim G(n_1/2, n_1 S_1/2)$;

   where the expressions of b_ϕ, B_ϕ, b^*, B^*, n_1 and S_1 were given there.

The novelty here is the assumption that C is unknown which requires the evaluation of its full conditional distribution. The density is given by

$$\pi(C|\beta_1, \beta_2, \phi) \propto f_N(\beta_1; X_2\beta_2, C^{-1}) f_W(C; n_W/2, n_W S_W/2)$$

$$\propto |C|^{1/2} \exp\left\{-\frac{1}{2}\mathrm{tr}[(\beta_1 - X_2\beta_2)(\beta_1 - X_2\beta_2)'C]\right\}$$

$$\times |C|^{(n_W - r + 1)/2} \exp\left\{-\frac{1}{2}\mathrm{tr}[n_W S_W C]\right\}$$

$$\propto |C|^{\frac{n_W - r}{2}} \exp\left\{-\frac{1}{2}\mathrm{tr}[(n_W S_W + (\beta_1 - X_2\beta_2)(\beta_1 - X_2\beta_2)')C]\right\}$$

which leads to

4. $C|\beta_1, \beta_2, \phi \sim W(n_W^*/2, n_W^* S_W^*/2)$
   where $n_W^* = n_W + 1$ and $n_W^* S_W^* = n_W S_W + (\beta_1 - X_2\beta_2)(\beta_1 - X_2\beta_2)'$.

A complete cycle of the Gibbs sampler involves successive sampling from the distributions given in steps 1-4. Generation from all these distributions is described in Chapter 1. Note that the blocks were naturally determined by the structure of the model.

Example 5.6 (continued) The model used is a special case of the above models and draws from the posterior distribution can be obtained. The application indicates convergence after 1500 iterations. Therefore, the remaining 2500 values from the two chains can be taken to form a sample of size 5000 from the posterior. The resulting histogram can be smoothed and the resulting marginal posteriors appear in Figure 5.7. They show a normal-like form for the populational growth and gamma-like form for the scale parameters with more variation in the population of the αs than for the βs.

Another important aspect is the assumed normality of errors at all levels. This is an unnecessary restriction now and in particular thicker-tailed distributions as the Student's t may be used. If the error distribution can be written as a (discrete or scale) mixture of normals, all full conditionals can be easily sampled from. In the context of exchangeable regression models, one may replace the first level equation by

$$\beta_i|\mu, \lambda_i, C \sim N(\mu, \lambda_i^{-1}C^{-1}) \text{ and } \lambda_i \sim F_\lambda \quad , \quad i = 1, ..., m$$

If F_λ is a Gamma distribution, the regression coefficients are t distributed. In this case, the full conditional distributions of $\beta = (\beta_1, ..., \beta_m)'$ and μ

alter only by the substitutions of C by $\lambda_i C$. The full conditional distribution of C now has $n_W^* S_W^* = n_W S_W + \Sigma_i \lambda_i (\beta_i - \beta_2)(\beta_i - \beta_2)'$. If $F_\lambda = G(n_\lambda/2, n_\lambda S_\lambda/2)$, the full conditional distribution of $\lambda = (\lambda_1, ..., \lambda_m)'$ is given by

$$\pi_\lambda(\lambda) \propto \prod_{i=1}^{m} f_N(\beta_i; \mu, \lambda_i^{-1} C^{-1}) \prod_{i=1}^{m} f_G(\lambda_i; n_\lambda/2, n_\lambda S_\lambda/2)$$

$$\propto \prod_{i=1}^{m} \lambda_i^{1/2} \exp\left\{-\frac{\lambda_i}{2}(\beta_i - \beta_2)' C(\beta_i - \beta_2)\right\} \lambda_i^{n_\lambda/2} \exp\left\{-\frac{\lambda_i}{2} n_\lambda S_\lambda\right\}$$

$$\propto \prod_{i=1}^{m} \lambda_i^{(n_\lambda+1)/2} \exp\left\{-\frac{\lambda_i}{2}[n_\lambda S_\lambda + (\beta_i - \beta_2)' C(\beta_i - \beta_2)]\right\}$$

and a posteriori the λ_i remain conditionally independent with distributions $G\{(n_\lambda + 1)/2, [n_\lambda S_\lambda + (\beta_i - \beta_2)' C(\beta_i - \beta_2)]/2\}$, $i = 1, ..., m$. So, minor modifications in the already existing steps and the introduction of an additional step with independent Gamma draws are the only changes required to robustify the model.

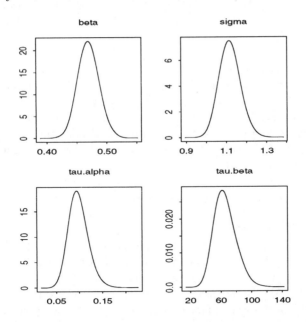

Figure 5.7. *Smoothed marginal density estimates for data on pregnant women.*

5.5.2 Dynamic models

The dynamic models considered here are given by (2.18)-(2.19) with constant variances of the observation and system disturbances, i.e., $\sigma_t^2 = \sigma^2$

and $W_t = W$, for all t. This restriction is aimed mainly at presentation clarity. It also provides for a more parsimonious model. The extension to the general case of unequal variances is not difficult and is left as an exercise.

As in hierarchical models, the specification of the model can be taken as a basis for blocking parameters. So, the natural choice is to form blocks $\beta_1, ..., \beta_n, \sigma^2$ and W.

The full conditional distributions of the β_t were obtained in section 2.5.2 and are given by $\pi(\beta_t|\beta_{-t}, \sigma^2, W) = N(b_t, B_t)$ where

$$
b_t = \begin{cases}
B_t(\sigma^2 F_t y_t + G'_{t+1} W^{-1} \beta_{t+1} + R^{-1} a) & , t = 1 \\
B_t(\sigma^2 F_t y_t + G'_{t+1} W^{-1} \beta_{t+1} + W^{-1} G_t \beta_{t-1}) & , t = 2, ..., n-1 \\
B_t(\sigma^2 F_t y_t + W^{-1} G_t \beta_{t-1}) & , t = n
\end{cases}
$$

and

$$
B_t = \begin{cases}
(\sigma^2 F_t F'_t + G'_{t+1} W^{-1} G_{t+1} + R^{-1})^{-1} & , t = 1 \\
(\sigma^2 F_t F'_t + G'_{t+1} W^{-1} G_{t+1} + W^{-1})^{-1} & , t = 2, ..., n-1 \\
(\sigma^2 F_t F'_t + W^{-1})^{-1} & , t = n
\end{cases}
$$

Assuming now independent priors $\phi = \sigma^{-2} \sim G(n_\sigma/2, n_\sigma S_\sigma/2)$ and $\Phi = W^{-1} \sim W(n_W/2, n_W S_W/2)$, their full conditional posterior distributions were also obtained in section 2.5.2 as $G(n_\sigma^*/2, n_\sigma^* S_\sigma^*/2)$ and $W(n_W^*/2, n_W^* S_W^*/2)$, respectively. Once again, both parameters are conditionally conjugate.

These full conditional distributions complete a cycle of the Gibbs sampler. They are all easy to sample from (see Chapter 1) and one can proceed with a sampling-based Bayesian inference. This approach was introduced by Carlin, Polson and Stoffer (1992). They also extended the analysis to models with non-normal disturbances ϵ_t and w_t and non-linear models. Non-normality was introduced through scale mixtures of normals and therefore the same methods used in section 5.5.1 can be applied here. Non-linearity was introduced with the replacement of the linear forms $F'_t \beta_t$ and $G_t \beta_{t-1}$ by arbitrary functions $F_t(\beta_t)$ and $G_t(\beta_{t-1})$. In these cases, the authors suggest the use of rejection methods with a normal density q based on the linear parts of the model.

Unfortunately, this approach may be very inefficient. The system equation introduces prior correlation between system parameters $\beta = (\beta_1, ..., \beta_n)'$. This correlation is controlled by the system variance matrix W. The smaller their elements, the larger in absolute value the correlation between the β_t will be. This correlation is partially preserved in the posterior although its quantification is more complicated. In the limit, when $W = 0$, the prior and consequently the posterior correlation is one. This is a highly-dimensional version of the same phenomenon depicted in Figure 5.2. The high dimensionality of the state space brings convergence problems to Gibbs sampling.

Typically, the values of W are much smaller than the values of σ^2. In this case, posterior correlation between state parameters will be high and the chain will tend to move slowly across the state space. As a result, a large number of iterations is required both for the burn-in period and for collecting the sample from the limiting distribution. In the latter case, it is advisable to retain draws from every kth iteration for a final sample with fixed size. The high chain autocorrelation will tend to force similar values at successive iterations and appropriate coverage of the parameter space will only be achieved with a large sample or with spacing between draws.

However, if the entries of the system variance matrix W are large then the correlation between the components of β will be low and the Gibbs sampler will work well. In this case, the system parameters experience large variation and the very use of dynamic models becomes questionable; little information will be passed through model parameters. Dynamic models should be used when there is relevant passage of information through the system and in this case the correlation between model parameters will be high.

There are two alternative approaches to high correlation, both described in section 5.3: reparametrization and block sampling. As previously discussed, it is generally preferable to sample correlated parameters in blocks when using Gibbs sampling. This is also possible for dynamic models by using equation (2.24). It shows that the full conditional distribution of the block β is normal and can be decomposed in tractable densities that can be obtained and sampled from the updating equations. Incorporating explicitly the conditional on σ^2 and W, each term in (2.24) is given by Bayes' theorem as

$$p(\beta_t|\beta_{t+1}, \sigma^2, W, y^t) \propto p(\beta_{t+1}|\beta_t, \sigma^2, W, y^t)p(\beta_t|\sigma^2, W, y^t)$$
$$\propto f_N(\beta_{t+1}; G_t\beta_t, W)f_N(\beta_t; m_t, C_t)$$

where the first term on the right hand side plays the role of the likelihood and the second term plays the role of the prior. From this perspective, the *observations* β_{t+1} form a regression model with design matrix G_t and parameters β_t whose *prior* is given by the updating equations. It becomes easy to obtain that

$$(\beta_t|\beta_{t+1}, \sigma^2, W, y^t) \sim N[(G_t'W^{-1}G_t + C_t^{-1})^{-1}(G_t'W^{-1}\beta_{t+1} + C_t^{-1}m_t),$$
$$(G_t'W^{-1}G_t + C_t^{-1})^{-1}] \tag{5.6}$$

for $t = 1, ..., n-1$ using results from section 2.5 (Exercise 2.18d). So, a scheme for sampling from the full conditional of the block β is given by:

1. Sample β_n from its updated distribution (2.22) and set $t = n - 1$.
2. Sample β_t from the distribution (5.6).
3. Decrease t to $t - 1$ and return to step 2 until $t = 1$.

Step 1 is obtained by running the Kalman filter from $t = 1$ to $t = n$ with given values of σ^2 and W. When running the filter, the updated means m_t and variances C_t, $t = 1, ..., n$, are stored for use in step 2.

The above sampling scheme draws a value from the full conditional $\pi(\beta|\sigma^2, W)$. It was independently proposed by Carter and Kohn (1994) and Fruhwirth-Schnatter (1994). Examples in these papers and in Shephard (1994) show that convergence becomes orders of magnitude faster than sampling each β_t at a time. Also, the computational cost of each iteration is higher but comparable in magnitude to the cost of obtaining and sampling from the full conditionals of the β_t.

The other alternative is reparametrization and was explored by Gamerman (1995). The main source of correlation between the β_t is induced by the system equation. On the other hand, the β_t can be completely determined by the values of β_1 and disturbances w_t. The system equation can be rewritten as $w_t = \beta_t - G_t\beta_{t-1}$, $t = 2, ..., n$. Setting $w_1 = \beta_1$ and applying recursively the system equation leads to the inverse relation

$$\beta_t = \sum_{l=1}^{t} \left(\prod_{k=1}^{t-l} G_{t-k+1} \right) w_l$$

for $t = 2, ..., n$ and $\beta_1 = w_1$. For the majority of dynamic models of interest, $G_t = G$, for all t. In this case, the above expression simplifies to

$$\beta_t = \sum_{l=1}^{t} G^{t-l} w_l$$

The model can be written in terms of the new parameters w_t and their full conditionals can be obtained (Exercise 5.14). The disturbances are prior independent by construction but are not posterior independent. So, this scheme is not as efficient as block sampling β. Nevertheless, it removes the main source of correlation, the system equation, from the sampling scheme. This ensures fast convergence of the sampling algorithm although in this case the computational cost of each iteration is higher. This reparametrization will be explored in the next chapter in connection with non-normal dynamic models.

5.6 Software: BUGS, CODA, gibbsit and itsim

One of the greatest impediments on the development of Bayesian inference was the difficulty of its implementation in practical situations. This difficulty was due to a host of possibilities for the specification of the prior distribution and to the difficulty of summarization of the resulting posterior distribution. The first source of difficulty is being eliminated by the introduction of symbolic languages that accommodate many specifications in a computational system. The second source of difficulty was greatly eliminated by the introduction of MCMC methods such as the Gibbs sampler that allow the analysis of complex models through decomposition and sampling from full conditional distributions.

Any system capable of specifying a variety of prior distributions for any given model and sampling from the resulting full conditionals would solve a great number of Bayesian problems. One such system is BUGS (Spiegelhalter, Thomas, Best and Gilks, 1995a), which stands as an acronym for Bayesian inference Using Gibbs Sampling. BUGS is a system developed at the Biostatistics Unit of the Medical Research Council, United Kingdom. It consists of a set of functions that allows specification of models and probability distributions for all its random components (observations and parameters). Model specification is surprisingly simple given the complexity of models that it can tackle. Among those models already analysed with BUGS and described in its manual (Spiegelhalter, Thomas, Best and Gilks, 1995b) are generalized linear models with random effects, regression analysis of survival data, models for spatially dependent data and non-parametric smoothing models.

For each combination of data set and model, BUGS outputs samples of model parameters at every $k \geq 1$ iterations after m iterations. The values of k and m as well as sampled parameters to be stored are chosen by the user. In addition, it provides sample-based estimates of posterior mean and credibility interval for the parameters. Both the system language as well as data input and output follow the syntax of the S language thus providing a useful interface for other data manipulations the user may wish to entertain. The system is freely available through *ftp* from *ftp.mrc-bsu.cam.ac.uk* in the directory *pub/methodology/bugs.*

The system recognizes conditional conjugacy and uses it to sample efficiently. Failing that, it uses rejection and adaptive rejection methods (section 1.5). This is perhaps the main drawback of the system. At the moment, it only samples from scalar parameters. So, for models with strong correlation structure as dynamic models, it should be used with care with respect to convergence. The latest release of BUGS at the time of writing, Version 0.5, already allows for specification of multivariate normal and Wishart distributions. Block sampling and graphical model specification are still unavailable but are planned for the next releases.

Example 5.7 The data in Figure 5.8 describe the evolution in the height of teeth of children through time and was introduced by Elston and Grizzle (1962). In addition to the dynamic component of the trajectories through time, there is also the hierarchical component of similarities between trajectories. Gamerman and Smith (1996) proposed a mixture model that incorporates both aspects above while still preserving the individuality of the series. Their model was given by

$$
\begin{aligned}
y_{ti} &\sim N(\theta_{ti}, \sigma^2) \\
\theta_{ti} &\sim (1-p)N(\theta_{t-1,i} + \lambda_{ti}, W_1) + pN(\mu_t, V_1) \\
\lambda_{ti} &\sim (1-p)N(\lambda_{t-1,i}, W_2) + pN(\gamma_t, V_2)
\end{aligned}
$$

$$\mu_t \sim N(\mu_{t-1} + \gamma_t, W_1)$$
$$\gamma_t \sim N(\gamma_{t-1}, W_2)$$

and was completed with independent vague priors μ_1 and $\gamma_1 \sim N(0, 10^3)$, σ^2, V_1, V_2, W_1 and $W_2 \sim IG(0.1, 0.1)$.

The analysis was performed using BUGS. Appendix 5.A below illustrates how the model is described in BUGS. It is written as in S language with the added bonus of probabilistic attributions. Note that discrete mixture of normals is not a common distribution and is not directly available. It is reproduced in BUGS using the same device of indicator variables described in section 1.3. For this data set, despite the dynamic nature of the model, a single long chain was used. The very short time length of the series reduces the problems caused by the correlation of successive state parameters. After discarding the first 10 000 iterations the next 1000 iterations constituted the sample used for inference. Part of the results is exhibited in Figure 5.9 where the mixture character of the model is evidenced.

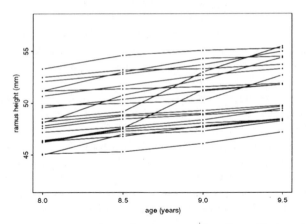

Figure 5.8. *Data on the height of the ramus tooth in boys. Each trajectory repre-sents the evolution of the height for one boy.*

Convergence diagnostics of a BUGS output can be performed through CODA (Best, Cowles and Vines, 1995). This is a system that may but does not need to be used in conjunction with BUGS and its name stands for an ana-gram of Convergence Diagnostics and Output Analysis. It is a collection of functions written in S-PLUS that is being maintained and distributed by the same research group responsible for BUGS. It contains many summarizing statistics and the convergence diagnostics of Gelman and Rubin (1992a), Geweke (1992) (also available in BUGS), Raftery and Lewis (1992) and Hei-delberger and Welch (1983). All simulations of Example 5.6 were made

with BUGS and convergence diagnostics including Figure 5.6 and Tables 5.1
and 5.2 were made with CODA.

Finally, the necessary calculations for the convergence prescription method
of Raftery and Lewis (1992) and the multiple chain analysis of Gelman
and Rubin (1992a) are freely available in the Statlib Library of statistical programs. They were written by the authors of the methods and
are respectively available in programs called gibbsit and itsim. They
can be retrieved by accessing the WWW site *http://lib.stat.cmu.edu* or by
sending an email message to *statlib@stat.cmu.edu* with a single line send
gibbsit from general or send itsim from S, respectively. More thorough descriptions of these and other programs for Bayesian inference can
be found in Carlin and Louis (1996).

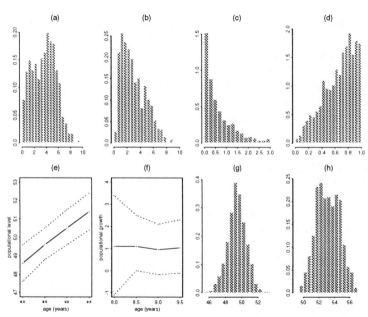

Figure 5.9. *Summary of estimation - posterior histograms of: (a) - σ; (b) - V₁;
(c) - W₂; (d) - p; (g) - θ₂,₁₂; (h) - θ₄,₁₈. Point estimates (full line) and 90%
credibility limits (broken lines) for populational parameters: (e)- μₜ; (f)- γₜ. Note
the difference between the histograms of individual levels in agreement (θ₂,₁₂) and
in disagreement (θ₄,₁₈) with the populational levels.*

Appendix 5.A Model specification for Example 5.6 in BUGS

```
model mixture;
const
m= 20, # number of series
n= 4; # number of times
```

```
var
y[n,m],theta[n,m],lambda[n,m],k,k1,mi1[n,m,2],mi2[n,m,2],vi1[n,2],vi2[n,2]
mu[n],gamma[n],eta[n],p,sigma,tau1,tau2,w1,w2,w0;
data in "ramus.dat";
inits in "ramus.in";
{
# Prior specification for hyperparameters
sigma ~ dgamma(0.1,0.1);
tau1 ~ dgamma(0.1,0.1);
tau2 ~ dgamma(0.1,0.1);
w1 ~ dgamma(0.1,0.1);
w2 ~ dgamma(0.1,0.1);
p ~ dbeta(1,1);
# Prior specification for 2nd. level parameters
w0 <- 0.001 ;
mu[1] ~ dnorm(0, w0 );
gamma[1] ~ dnorm(0, w0);
# Prior specification for 1st. level parameters
k ~ dbern( p );
k1 <- 1 + k;
vi1[1,1] <- w0;
vi1[1,2] <- tau1;
vi2[1,1] <- w0;
vi2[1,2] <- tau2;
for(i in 1:m) {
mi1[1,i,1] <- 0;
mi1[1,i,2] <- mu[1];
mi2[1,i,1] <- 0 ;
mi2[1,i,2] <- gamma[1];
theta[1,i] ~ dnorm( mi1[1,i,k1] , vi1[1,k1] );
lambda[1,i] ~ dnorm( mi2[1,i,k1] , vi2[1,k1] );
# Observation equation
y[1,i] ~ dnorm(theta[1,i],sigma) }
# Model specification for t=2, ... , n
# Evolution for 2nd. level parameters
for(t in 2:n) {
gamma[t] ~ dnorm( gamma[t-1], w2 );
eta[t] <- mu[t-1] + gamma[t];
mu[t] ~ dnorm(eta[t], w1);
# Evolution for 1st. level parameters
vi1[t,1] <- w1;
vi1[t,2] <- tau1;
vi2[t,1] <- w2;
vi2[t,2] <- tau2;
```

```
for(i in 1:m) {
mi2[t,i,1] <- lambda[t-1,i];
mi2[t,i,2] <- gamma[t];
lambda[t,i] ~ dnorm( mi2[t,i,k1] , vi2[t,k1] );
mi1[t,i,1] <- beta[t-1,i]+lambda[t,i];
mi1[t,i,2] <- mu[t];
theta[t,i] ~ dnorm( mi1[t,i,k1] , vi1[t,k1] );
# Observation equation
y[t,i] ~ dnorm( theta[t,i],sigma) } }
}
```

5.7 Exercises

1. Consider the Gibbs sampler with transition kernel (5.2).

   (a) Show that the chain with a complete scan over all components is not reversible.

   (b) Under what conditions is the chain with transitions formed by individual changes on a single component reversible?

   (c) Show that the chain that takes each iteration to consist of the complete scan through the components followed by another scan through the components in reversed order is reversible (Besag, 1986).

2. (Casella and George, 1992) Let x and y be random quantities with conditional densities $f(x|y) = ye^{-yx}$, $x > 0$ and $f(y|x) = xe^{-xy}$, $y > 0$. Show that the only possible solution for $f(x)$ is $f(x) = 1/x$, which is not a proper density, and that Gibbs sampling cannot be applied in this case.

3. Consider Example 5.1 but assume now that the regions A and B have non-null intersection for only one of the axes, for example, the θ_1 axis. Discuss whether conditions for convergence to the joint distribution are satisfied.

4. (MacEachern and Berliner, 1994; O'Hagan, 1994) Consider the estimation of a real function $\psi = t(\theta)$ from a stream of n succesive values from a chain. Form k sub-samples of size $m = n/k$ by skipping every k iterations and assume that k is large enough to ensure approximate independence between sample values. Denote by $\hat{\psi}_1, ..., \hat{\psi}_k$ the averages of the sub-samples and $\hat{\psi} = (1/k)\sum_{j=1}^{k} \hat{\psi}_j$ the average over the complete sample.

   (a) Show that $Var(\hat{\psi}_j) = Var(\psi)/m$, for all j.

   (b) Use the Cauchy-Schwarz (or correlation) inequality $Cov(x,y) \leq \sqrt{Var(x)Var(y)}$ to show that $Var(\hat{\psi}) \leq Var(\hat{\psi}_j)$.

5. Consider the simple linear regression model $y_i = \beta_1 + \beta_2 x_i + e_i$ where $e_i \sim N(0, \sigma^2)$ are independent, $i = 1, ..., n$, with non-informative marginal $p(\beta) \propto k$.

   (a) Show that if α has components $\alpha_1 = \beta_1 + \beta_2 \bar{x}$ and $\alpha_2 = \beta_2$ then α_1 and α_2 are conditionally independent a posteriori given σ^2.

   (b) Show that using α is equivalent to centering the covariate and using the model $y_i = \alpha_1 + \alpha_2(x_i - \bar{x}) + e_i$.

   (c) Generalize the result to multiple linear regression.

6. Show that for the random effects model of Example 5.4, posterior correlations are given by

$$Cor(\mu, \alpha_i) = -\left(1 + \frac{\sigma^2/n_i}{\tau^2/m}\right)^{-1/2} \quad \text{and} \quad Cor(\alpha_i, \alpha_j) = \left(1 + \frac{\sigma^2/n_i}{\tau^2/m}\right)^{-1}$$

   in the original parametrization,

$$Cor(\mu, \beta_i) = -\left(1 + \frac{m\tau^2}{\sigma^2/n_i}\right)^{-1/2} \quad \text{and} \quad Cor(\beta_i, \beta_j) = \left(1 + \frac{m\tau^2}{\sigma^2/n_i}\right)^{-1}$$

   in the centered parametrization and

$$Cor(\nu, \xi_i) = 0 \quad \text{and} \quad Cor(\xi_i, \xi_j) = -\frac{1}{m}$$

   in the parametrization of Vines, Gilks and Wild (1994).

7. (O'Hagan, 1994) Consider the situation of Example 5.5 where $\theta = (\theta_1, \theta_2)$ is bivariate and $\pi(\theta)$ is given by the table of probabilities below

		θ_2	
		0	1
θ_1	0	$p/2$	$(1-p)/2$
	1	$(1-p)/2$	$p/2$

   (a) Obtain that $\pi(\theta_i) = bern(1/2)$, $i = 1, 2$, and that the posterior correlation between θ_1 and θ_2 is $\rho = 2p - 1$.

   (b) Show that $Pr(\theta_1^{(j)} = 1 | \theta_1^{(j-1)} = 1) = Pr(\theta_1^{(j)} = 0 | \theta_1^{(j-1)} = 0) = p^2 + (1-p)^2$ and, consequently, $Pr(\theta_1^{(j)} = 1 | \theta_1^{(j-1)} = 0) = Pr(\theta_1^{(j)} = 0 | \theta_1^{(j-1)} = 1) = 2p(1-p)$.

   (c) Show that if $p_j = Pr(\theta_1^{(j)} = 1)$, $j = 0, 1, ...$, then $p_j = \rho^2 p_{j-1} + b$ where $b = 2p(1-p)$ and derive that $p_j = \rho^{2(j+1)} p_0 + b(1 - \rho^{2(j+1)})/(1 - \rho^2)$.

   (d) Show that the transition matrix formed by the marginal chain $(\theta_1^{(j)})_{j \geq 0}$ has eigenvalues 1 and ρ.

(e) Show that in the limit, $p_j \to b/(1 - \rho^2) = 1/2$.

(f) Plot $p_j \times j$ for $p = 0.999$ for a given value of p_0.

8. Consider the chain $Z_k^{(j)}$ described in section 5.4 and its transition matrix

$$P_k = \left(\begin{array}{cc} 1 - \alpha & \alpha \\ \beta & 1 - \beta \end{array} \right)$$

(a) Show that the equilibrium distribution is $(\pi_0, \pi_1) = (\beta, \alpha)/(\alpha + \beta)$ and that

$$P_k^l = \left(\begin{array}{cc} \pi_0 & \pi_1 \\ \pi_0 & \pi_1 \end{array} \right) + \frac{\lambda^l}{\alpha + \beta} \left(\begin{array}{cc} \alpha & -\alpha \\ -\beta & \beta \end{array} \right)$$

where $\lambda = 1 - \alpha - \beta$.

(b) Show that if $Pr(Z_k^{(m)} = i | Z_k^{(0)} = j)$ is required to be within ϵ from its limiting value π_i, i.e., $|Pr(Z_k^{(m)} = i | Z_k^{(0)} = j) - \pi_i| \leq \epsilon$, $i, j = 0, 1$, then

$$m \geq m^* = \frac{log\left[\frac{(\alpha + \beta)\epsilon}{max(\alpha, \beta)} \right]}{log |\lambda|}$$

and therefore $m = m^* k$ should be taken as the burn-in period.

9. Still in the conditions of section 5.4.2, show

(a) using (4.9) that for large n

$$\bar{Z}_{k,n} \dot\sim N\left[q, \frac{1}{n} \frac{(2 - \alpha - \beta)\alpha\beta}{(\alpha + \beta)^3} \right]$$

(b) that the smallest n satisfying $Pr(|\bar{Z}_{k,n} - q| < r) = s$ is

$$n^* = \frac{(2 - \alpha - \beta)\alpha\beta}{(\alpha + \beta)^3} \left\{ \frac{z_{(1+s)/2}}{r} \right\}^2$$

where z_γ is the γ quantile of the $N(0, 1)$ distribution.

(c) that n^* is minimized in the case of independent sampling in which case $1 - \alpha = \beta = q$ and is given by

$$n^* = \frac{q(1 - q)z_{(1+s)/2}^2}{r^2}$$

10. Consider m parallel chains and a real function $\psi = t(\theta)$. There are m trajectories $\{\psi_i^{(1)}, \psi_i^{(2)}, ..., \psi_i^{(n)}\}$, $i = 1, ..., m$, for ψ. The variances between chains B and within chains W are given by

$$B = \frac{n}{m - 1} \sum_{i=1}^m (\bar\psi_i - \bar\psi)^2 \text{ and } W = \frac{1}{m(n - 1)} \sum_{i=1}^m \sum_{j=1}^n (\psi_i^{(j)} - \bar\psi_i)^2$$

where $\bar\psi_i$ is the average of observations of chain i, $i = 1, ..., m$, and $\bar\psi$ is the average of these averages.

(a) Use the ergodic theorem to show that W, B and $\hat{\sigma}_\phi^2 = (1 - 1/n)W + (1/n)B \to \sigma_\psi^2$, the variance of ψ, when $n \to \infty$.

(b) (Gelman and Rubin, 1992a) Let $V = \sigma_\psi^2 + Var(\bar{\psi})$ and $\hat{V} = \hat{\sigma}_\psi^2 + B/mn$ be an estimator of V. Show that $E(\hat{V}) = V$ and if the distribution of $\hat{V}/V$ is approximated by a χ_ν^2 then ν is estimated by the method of moments as

$$\hat{\nu} = 2\frac{\hat{V}^2}{\widehat{Var}(\hat{V})}$$

Note: the potential scale reduction estimator originally proposed was given by $\hat{R} = \sqrt{(\hat{V}/W)(\hat{\nu}/(\hat{\nu} - 2))}$.

11. (Zellner and Min, 1995) Derive the central limit theorem for the Rao-Blackwellized estimator of marginal densities given by (5.5).

    (a) Derive confidence intervals and a test for convergence based on the result and using the difference criterium statistic $\hat{\eta}$ evaluated at m different states $\theta_1, ..., \theta_m$.

    (b) Assuming a vague prior for η, obtain its posterior distribution and construct credibility intervals for η.

    (c) Repeat items (a) and (b) to derive tests and confidence and credibility intervals for convergence and for correct convergence based on the ratio criterium statistics.

12. Specify a version of the 3-stage hierarchical model and obtain the full conditional distributions required for implementation of the Gibbs sampler.

13. Obtain the full conditional distributions required for implementation of the Gibbs sampler for the dynamic model with observational and system variances having independent prior distributions $IG(n_{\sigma,t}/2, n_{\sigma,t}S_{\sigma,t}/2)$ and $IW(n_{W,t}/2, n_{W,t}S_{W,t}/2)$ respectively, $t = 1, ..., n$. Compare and discuss in this context the relative efficiency of the methods based on separate sampling of the β_t, block sampling of β and sampling via the reparametrizations w_t.

14. Show that

$$\beta_t = \sum_{l=1}^{t} \left(\prod_{k=1}^{t-l} G_{t-k+1} \right) w_l$$

for $t = 2, ..., n$ and $\beta_1 = w_1$ and, if $G_t = G$, for all t, the above expression simplifies to

$$\beta_t = \sum_{l=1}^{t} G^{t-l} w_l$$

Show also that the full conditional distribution of w_t is $N(b_t, B_t)$ where

$$b_t = B_t \sum_{l=t}^{n} V^{-1} H_{tl}(y_l - k_{tl}) \text{ and } B_t^{-1} = W^{-1} + \sum_{l=t}^{n} V^{-1} H_{tl} H'_{tl}$$

for $t = 2, ..., n$ and

$$b_1 = B_1 \left[R^{-1} a + \sum_{l=1}^{n} V^{-1} H_{1l}(y_l - k_{1l}) \right] \text{ and } B_1^{-1} = R^{-1} + \sum_{l=1}^{n} V^{-1} H_{1l} H'_{1l}$$

where $H'_{tl} = F'_t G^{l-t}$ and $k_{tl} = F'_t \sum_{i=1, i \neq t}^{j} G^{l-i} w_i$, $l \geq t$, $t = 1, ..., n$.

## CHAPTER 6

Metropolis-Hastings algorithms

6.1 Introduction

In this chapter, Markov chains known under a generic name of Metropolis-Hastings will be presented and discussed. This name stems from papers by Metropolis et al. (1953) and Hastings (1970). These are considered as basic papers for the characterization of the method, although other papers including Barker (1965) and Peskun (1973) have also brought relevant contributions to the method.

The original paper by Metropolis et al. (1953) deals with the calculation of properties of chemical substances and was published in the *Journal of Chemical Physics*. Nevertheless, it later proved itself to have a great impact in Statistics and Simulation.

Consider a substance with d molecules positioned at $\theta = (\theta_1, ..., \theta_d)'$. In this case, the component θ_i is formed by the bidimensional vector of positions in the plane of the ith molecule. From Statistical Mechanics, the density of these positions is given by (5.1) where V is the potential between molecules. The potential energy of the substance is then given by $E(\theta) = \Sigma_{i,j} V(\theta_i, \theta_j)/2$.

The calculation of the equilibrium value of any chemical property is given by the expected value of this property with respect to the distribution of the vector of positions. Direct calculation of the expectation is not feasible for d large and is replaced by a Monte Carlo estimate. Metropolis et al. (1953) suggested the following method to deal with the difficult problem of sampling from this density:

1. Start with any initial configuration $\theta^{(0)} = (\theta_1^{(0)}, ..., \theta_d^{(0)})'$ and set the iteration counter $j = 1$.

2. Move the particles from previous positions $\theta^{(j-1)} = (\theta_1^{(j-1)}, ..., \theta_d^{(j-1)})'$ according to a uniform distribution centered at these positions in order to obtain new positions $\phi = (\phi_1, ..., \phi_d)'$.

3. Calculate the change ΔE in the potential energy caused by the move. The move in step 2 is accepted with probability $\min\{1, e^{-c\Delta E}\}$. If the move is accepted, $\theta^{(j)} = \phi$. Otherwise, $\theta^{(j)} = \theta^{(j-1)}$.

4. Change the counter from j to $j+1$ and return to step 2 until convergence is reached.

After convergence, the vector of positions generated by the method has distribution with density (5.1). It is evident that the above method defines a Markov chain as the transitions depend only on the positions at the previous stage. However, it is not obvious that the method converges to a equilibrium distribution and that this distribution is given by (5.1). Metropolis et al. (1953) present a heuristic proof of this result. The same proof is valid for the case where the moves to ϕ are made according to any symmetric distribution centered at previous positions. This defines a transition kernel q that depends on (θ, ϕ) through $|\phi-\theta|$. Hastings (1970) refers to the above algorithm in this extended form as the Metropolis method. In the next section, a more general version of the algorithm and the proof of its convergence will be presented.

Note that the above algorithm includes an additional step that was not present in the chains previously presented. The transition mechanism now depends on a proposed transition q *and* a subsequent step of evaluation of this proposal. Note that the proposed positions are completely unrelated from the equilibrium distribution but this is represented in the overall transition through the acceptance probability because

$$\frac{\pi(\phi)}{\pi(\theta^{(j-1)})} = \frac{\exp\{-E(\phi)/kT\}}{\exp\{-cE(\theta^{(j-1)})/kT\}} = \exp\{-c\Delta E\}$$

Another important point is that the resulting chain may remain in a low energy (or equivalently, high density) position for many iterations. In this case, it is likely that the proposal will lead to very high energy (very low density) points and $\Delta E \gg 0$ forcing an acceptance probability very close to 0. Computationally, this is not desirable and transition kernels must be carefully chosen to avoid such low acceptance rates.

The next section presents a more common and complete version of the algorithm following the work of Hastings (1970). Important special cases are presented in section 6.3 and in section 6.4 variations of the method are discussed. These variations include blocking and the relationship with Gibbs sampling. Finally, application of the algorithm to the context of generalized linear models with hierarchical and dynamic structure is discussed. A very nice expository introduction to Metropolis-Hastings algorithms is also provided by Chib and Greenberg (1995).

6.2 Definition and properties

Consider a distribution π from which a sample must be drawn via Markov chains. Again, it is worth stressing that this task will only make sense if the non-iterative generation of π is very complicated or expensive. In this case, a transition kernel $p(\theta, \phi)$ must be constructed in a way such that π is the equilibrium distribution of the chain. A simple way to do this is to consider reversible chains where the kernel p satisfies

$$\pi(\theta)p(\theta,\phi) = \pi(\phi)p(\phi,\theta)\,,\;\forall\,(\theta,\phi) \tag{6.1}$$

As previously seen in section 4.6, this is the reversibility condition of the chain. Equation (6.1) is also referred to as the detailed balance equation. Even though this is not a necessary condition for convergence, it is a sufficient condition in order that π be the equilibrium distribution of the chain.

The kernel $p(\theta,\phi)$ consists of 2 elements: an arbitrary transition kernel $q(\theta,\phi)$ and a probability $\alpha(\theta,\phi)$ such that

$$p(\theta,\phi) = q(\theta,\phi)\alpha(\theta,\phi)\,,\;\text{if } \theta \neq \phi$$

So, the transition kernel defines a density $p(\theta,\cdot)$ for every possible value of the parameter different from θ. Consequently, there is a positive probability left for the chain to remain at θ given by

$$p(\theta,\theta) = 1 - \int q(\theta,\phi)\alpha(\theta,\phi)d\phi$$

These two forms can be grouped in the general expression

$$p(\theta,A) = \int_A q(\theta,\phi)\alpha(\theta,\phi)d\phi + I(\theta \in A)\left[1 - \int q(\theta,\phi)\alpha(\theta,\phi)d\phi\right] \tag{6.2}$$

for any subset A of the parameter space. So, the transition kernel defines a mixed distribution for the new state ϕ of the chain. For $\phi \neq \theta$, this distribution has a density and for $\phi = \theta$, this distribution has a probability atom.

Hastings (1970) proposes to define the acceptance probability in such a way that when combined with the arbitrary transition kernel, it defines a reversible chain. The expression most commonly cited for the acceptance probability is

$$\alpha(\theta,\phi) = \min\left\{1, \frac{\pi(\phi)q(\phi,\theta)}{\pi(\theta)q(\theta,\phi)}\right\} \tag{6.3}$$

Algorithms based on chains with transition kernel (6.2) and acceptance probability (6.3) will be referred to as Metropolis-Hastings algorithms from now on. This is an acknowledgement of the importance of the contribution from both papers. Hastings (1970) refers to the ratio appearing in (6.3) as the test ratio. A more general expression for α including (6.3) and the acceptance probability used by Barker (1965) as special cases is presented in Hastings (1970). Optimality of these choices can be discussed in terms of minimization of asymptotic variance of moment estimates. Peskun (1973) shows for the discrete case that (6.3) is optimal in a large class of choices. He also shows that for suitable choices of the proposal transition q, Markov chain sampling can be more precise than independent sampling.

The proof that (6.1) is satisfied by p given in (6.2) and hence defines a reversible chain with equilibrium distribution π follows directly from (6.3) and is left as an exercise. Note that (6.1) is satisfied by p but not by q. The proposal transition kernel q has up to now been kept arbitrary and is thus a flexible tool for the construction of the algorithm. Roberts and Smith (1994) show that if q is irreducible and aperiodic and $\alpha(\theta, \phi) > 0$, for every possible value of (θ, ϕ), then the algorithm defines an irreducible and aperiodic chain with transition kernel p given by (6.2) and limiting distribution π.

In practical terms, simulation of a draw from π using the Markov chain defined by the transition (6.2) can be set up as follows:

1. Initialize the iteration counter $j = 1$ and set an arbitrary initial value $\theta^{(0)}$.

2. Move the chain to a new value ϕ generated from the density $q(\theta^{(j-1)}, \cdot)$.

3. Evaluate the acceptance probability of the move $\alpha(\theta^{(j-1)}, \phi)$ given by (6.3). If the move is accepted, $\theta^{(j)} = \phi$. If it is not accepted, $\theta^{(j)} = \theta^{(j-1)}$ and the chain does not move.

4. change the counter from j to $j+1$ and return to step 2 until convergence is reached.

Step 3 is performed after the generation of an independent uniform quantity u. If $u \leq \alpha$, the move is accepted and if $u > \alpha$ the move is not allowed. The transition kernel q defines only a possible move that can be confirmed according to the value of α. For that reason, q is generally referred to as the proposal kernel or proposal (conditional) density when looked upon as a (conditional) density $q(\theta, \cdot)$. Other terms sometimes used are probing kernel or density.

In any of the forms of the Metropolis algorithm, q defines a symmetric transition around the previous positions of the molecules. Therefore, $q(\theta, \phi) = q(\phi, \theta)$, for every (θ, ϕ) and the acceptance probability becomes

$$\alpha(\theta, \phi) = \min\left\{1, \frac{\pi(\phi)}{\pi(\theta)}\right\}$$

depending only on a simplified test ratio $\pi(\phi)/\pi(\theta)$, the ratio of the posterior density values at the proposed and previous positions of the chain.

Note also that the chain may remain in the same state for many iterations. A useful monitoring device of the method is given by the average percentage of iterations for which moves are accepted. Hastings (1970) suggests that this acceptance rate should always be computed in practical applications.

The success of the method depends on not having a very low acceptance rate. A naive approach to the problem is to make the chain move very slowly, i.e., to drive the chain so that its displacements are minute. Assuming for simplicity that $q(\cdot, \cdot)$ and $\pi(\cdot)$ are continuous, similar values for previous and proposed states will lead to a test ratio and hence acceptance probability close to 1. Following this strategy, the chain will have very high acceptance rates and most proposed moves are accepted. The chain however must be capable to traverse the whole parameter space in order to converge to the equilibrium distribution. Very small moves will make it take many iterations to converge. On the other hand, large displacements may be proposed but they are likely to fall in the tails of the posterior distribution causing a very low value for the test ratio. The chain moves, determined by q, must be paced in such a way as to provide considerable displacements from the current state but with substantial probability, determined by α, of being accepted.

It is also crucial that the proposal kernels are easy to draw from as the method replaces the difficult generation of π by many generations proposed from q. Another less obvious but equally important requirement to be met by q is the correct tuning of the moves it proposes to ensure that moves covering the parameter space can be made and accepted in real computing time.

Optimization studies in this area are not conclusive and are likely never to be. The diversity of models that can be treated and of transitions q that can be proposed make it extremely difficult to allow for general results. Current reasoning, expressed in an applied context in Bennett, Racine-Poon and Wakefield (1995), Besag et al. (1995) and other authors seem to indicate to the direction of acceptance rates between 20% to 50% In a specific theoretical context, Gelman, Roberts and Gilks (1996) obtained optimal acceptance rates of 24% for high-dimensional problems with normal densities π and q. These values should be looked at only as a generic indication rule and never as a compulsory determination. In the final section, an application with two sampling schemes with very high acceptance rates (larger than 90 %) is shown. The performance of the schemes are very different with one of them showing fast convergence whereas the second one has a very slow convergence.

The test ratio can be rewritten as

$$\frac{\pi(\phi)/q(\theta, \phi)}{\pi(\theta)/q(\phi, \theta)} \tag{6.4}$$

Acceptance of proposed values is based on the ratio of target and proposed density. So, there is a connection here with the resampling schemes described in section 1.5. There, the proposal density q was to be chosen as similar as possible to π to increase acceptance rates but the methods were not iterative. Also, for the rejection method, the rejection probability depended only on the numerator in (6.4).

The target distribution π enters the algorithm through the test ratio in the form of the ratio $\pi(\phi)/\pi(\theta)$, as in the resampling methods. So again, the complete knowledge of π is not required. In particular, proportionality constants are not needed. When π is a posterior density, even though its functional form is always known, the value of the proportionality constant is rarely known. So, the algorithm is particularly useful for applications to Bayesian inference.

Many of the comments made about Gibbs sampling in the previous chapter are also valid for the Metropolis-Hastings algorithm. So, the discussion about single long against multiple chains is just as relevant here. In using a single long chain, particular attention must be given to spacing between values of the chain taken for the resulting sample. The possible repetition of the same state for many iterations does not hinder convergence properties but makes it more common to produce sequences of repeated values, even after some spacing is allowed. A sample must adequately cover the complete parameter space so it is important that its values are not unnecessarily influenced by their predecessors.

Formal and informal convergence techniques described in Chapter 5 can all be used here. The exception is made up of those based on complete knowledge of conditional densities. Typically, but not necessarily, Metropolis-Hastings algorithms are used when these are not completely known and hence difficult to sample from. When the complete conditional densities are known, Gibbs sampling is generally used. Besag et al. (1995) argue against taking this approach as a general rule. They reason that Gibbs sampling does not take into account the previous value of the component being updated and is therefore restrictive. Questions relative to optimization of the algorithm through reparametrization or blocking are deferred to section 6.4.

6.3 Special cases

As described in the previous section, there is total flexibility for the choice of the proposal transition q apart from a few technical restrictions. Some general considerations have already been made and now we turn to some specific classes. It should be pointed out that although a chain is defined by its transition kernel p and not by a proposal transition q, the names used to categorize the algorithm generally refer to properties of q rather than p.

6.3.1 Symmetric chains

A chain is said to be symmetric if its transition kernel p is symmetric in its arguments, namely $p(\theta, \phi) = p(\phi, \theta)$, for every pair (θ, ϕ) of states. For the Metropolis-Hastings algorithms, the notion of symmetric chain is applied to the proposed transition q. An example of a symmetric chain is the Metropolis version of the algorithm. If q depends on (θ, ϕ) only through $|\phi - \theta|$ then $q(\theta, \phi) = q(\phi, \theta)$. In this case, the acceptance probability reduces to $\min\{1, \pi(\phi)/\pi(\theta)\}$ and does not depend on q. A computational simplification that may well prove to be substantial is thus obtained.

6.3.2 Random walk chains

Again, this characterization refers to the proposal transition q. From Chapter 4, we know that a random walk is a Markov chain with evolution given by $\theta^{(j)} = \theta^{(j-1)} + w_j$ where w_j is a random variable with distribution independent of the chain. In general, the disturbances w_j are independent and identically distributed with density f_w. The chain has proposed moves according to $q(\theta, \phi) = f_w(\phi - \theta)$. If f_w is symmetric around 0, the chain is symmetric and all comments above are valid here. The Metropolis algorithm can then be seen as a special case of a random walk chain.

This is a very common option and most practical implementations of Metropolis-Hastings algorithms use this scheme. The most used choices for f_w are the normal (Muller, 1991b) and Student's t (Geweke, 1992) distributions centered at the origin. Proposed values are then based around the previous values of the chain. An important point still remaining is the choice of the dispersion of f_w. Large values for the variance allow moves that are very distant from previous values but at the likely cost of very small acceptance rates. On the other hand, small values for the variance only allow moves close to the previous values but with high acceptance rates. Tierney (1994) suggests setting the variance matrix of f_w as cV where c is a multiplying scalar playing the role of a tuning constant and V is some form of approximation for the posterior variance (see Chapter 3). This allows the moves along the components of θ to be of the same size relative to the spread of the posterior distribution. The choice of the tuning constant depends on the form of optimization desired (high acceptance rates/large moves). Metropolis et al. (1953) discuss this issue in the context of their application. Tierney (1994) suggests values between $1/2$ and 1. Bennett, Racine-Poon and Wakefield (1995) report the use of a variety of values in the context of non-linear hierarchical models and also obtain the same recommendation based on sample sizes prescribed by the technique of Raftery and Lewis (1992). Gelman, Roberts and Gilks (1996) obtained their optimal acceptance rates for normal random walk proposals with c between 2 and 3.

6.3.3 Independence chains

In this case, the proposed transition is formulated independently of the previous position θ of the chain. So, $q(\theta, \phi) = f(\phi)$. It may seem that the independence from the previous state disagrees with the Markovian property of the chain. Once again, it is worth remembering that q is just a proposal that is combined with an acceptance probability α to give the transition kernel p of the algorithm. This transition depends on the previous state, preserving the Markovian structure.

Using expression (6.4) for the test ratio, it reduces to $w(\phi)/w(\theta)$ where $w = \pi/f$. The weight function works like the weight function in weighted resampling methods (section 1.5.2). One popular choice for f is the prior density as in section 3.5.2 (West, 1996; Knorr-Held, 1997). In this case, $w = l$, the likelihood function and the acceptance probability is $\alpha(\theta, \phi) = \min\{1, l(\phi)/l(\theta)\}$.

The use of the prior distribution as the basis for a resampling scheme was discussed in section 3.5 with advantages and disadvantages equally relevant here. Computationally, it has the advantages of producing one of the simplest expressions for α. The main disadvantage is when there is conflict between prior and likelihood information. Values more likely to be sampled are those supported by the prior and they will have little posterior support. Values highly supported by the likelihood (and probably by the posterior) will have very little chance of being drawn by such a scheme.

Proposal densities incorporating the likelihood help to avoid this situation. Normal approximations to the likelihood were used by Bennett, Racine-Poon and Wakefield (1995) and Chib and Greenberg (1994) to form normal independence proposals with and without combination with the prior distribution. Jacquier, Polson and Rossi (1994) use moment matching approximation to construct a Gamma proposal in the context of stochastic volatility models.

The general rule for independence chains is to avoid large variation in the weight function as this increases the chances of a chain being retained for many iterations in states with large weights. So, it is recommended that f is chosen in order to make the function w as constant as possible, or at least bounded. As f and π are both densities, this is equivalent to recommending that f and π are as similar as possible. That rules out the use of a prior proposal in case of disagreement between prior and likelihood.

Tierney (1994) suggests avoiding densities f with thin tails such as the normal distribution and to use instead t densities with small numbers of degrees of freedom. In this way, the weight function will not be so strongly affected by the tails of f. By doing that, the weight function becomes less likely to have large variations.

6.3.4 Other forms

Tierney (1994) presents many other classes of proposal transitions that may be used. In particular, he discusses the use of the rejection method in independence chains. In the case of sampling from π based on rejection sampling from f, f should be an envelope function for π. This requires finding a constant A to achieve this task, which can be difficult due to the complicated form of π. Large values of A ensure a proper envelope at the expense of high rejection rates and small values of A may not ensure a complete envelope. This structure can be put in the context of the Metropolis-Hastings algorithm without having to enforce a (possibly difficult to obtain) specification of the envelope constant A (Exercise 6.6). Chib and Greenberg (1995) give the details of this application of rejection methods in the Metropolis-Hastings context.

Gilks, Best and Tan (1995) generalized the adaptive rejection sampling scheme to non-log-concave densitites using the same idea. The approximating piecewise exponential density does not provide an envelope for these densities. The rejection step may then be replaced by a Metropolis-Hastings step just as outlined above. Limiting results of Metropolis-Hastings algorithms ensure that sampling is still correct.

Another possibility is the extension of random walk chains to autoregressive chains where $\theta^{(j)} = a + b\theta^{(j-1)} + w_j$ (see Example 6.2 below). Taking the value of $b = 1$ reduces to random walk chains with incorporation of the constant a to the distribution f_w. Taking $b = -1$ produces alternations in the chain forcing negative autocorrelations. These two choices of b are compared for the univariate normal distribution by Hastings (1970) and for a bivariate normal distribution with high correlation by Chib and Greenberg (1995). In both examples, the alternating chain produces better estimates of the target distribution. Barone and Frigessi (1989) provide further theoretical support for this choice of value of b. This alternating effect is not new in simulation and is generally known as antithetic variables with good properties in terms of reducing variance of estimates (Ripley, 1987).

The algorithm of Ritter and Tanner (1992) presented in the previous chapter is, in fact, based on an approximation to the target distribution. This can be improved upon by incorporating it into a Metropolis-Hastings algorithm as the proposal kernel. If the kernel admits a density then this proposal density will appear in the expression of the acceptance probability.

Finally, further proposal densities that do not fit into any of the above schemes will be presented in the application section. These proposals were motivated by the structure of the model and in similar inferential procedures already available. The diversity of options presented in this section serves the purpose of showing the vast field available for exploration when simulating via Markov chains. Most of them have only started to be explored.

6.4 Hybrid algorithms

In this chapter, a simulation scheme using Markov chains called the Metropolis-Hastings algorithm is being presented. This scheme was introduced in a general form in section 6.2 and some special cases were presented in section 6.3. The aim of this section is to present some of the capabilities of the scheme especially in connection with componentwise transition and combinations of different transition schemes.

In the previous chapter, a scheme based on transition by components was presented, the Gibbs sampler. The similarity between the componentwise sampling schemes will be clarified and schemes combining the two algorithms are presented. Also in this section, a discussion about blocking and reparametrization is made.

6.4.1 Componentwise transition

The previous section showed some of the possibilities that are available with Metropolis-Hastings algorithms. In all cases presented, the quantity of interest for sampling θ was updated in a single block. Again here, new transition forms are available when the components $\theta_1, ..., \theta_d$ of θ are used separately. Hastings (1970) and Tierney (1994) discuss this possibility. In particular, the components of θ can be updated or changed in the following forms:

a) At each iteration, a single component is updated and the choice of the component is made at random between the d components.

b) At each iteration, a single component is updated and the choice of the component is made in a fixed pre-specified order of the d components. For example, the components are updated in the order $1 \rightarrow 2 \rightarrow \cdots \rightarrow d$.

The above forms are examples of mixtures in case (a) and cycles in case (b) of transitions. In the first case, define transitions p_m with a common equilibrium distribution π and probabilities or weights w_m, $m = 1, ..., r$, satisfying $w_m \geq 0$ and $\sum_{m=1}^{r} w_m = 1$. A mixture transition p is formed by taking $p = \sum_{m=1}^{r} w_m p_m$. Case (a) above is a special case of a mixture with $r = d$, $w_m = 1/d$ and each transition p_m moves only the mth component of θ, $m = 1, ..., d$.

Properties of the transitions p_m are passed on to the mixture transition p. First of all, the mixture kernel p defines a transition kernel of a Markov chain with equilibrium distribution π. Also, if one of the component transition kernels is irreducible and aperiodic then the mixture kernel is irreducible and aperiodic.

In the case of cycle transition kernels with component transition kernels p_c, $c = 1, ..., r$, then an iteration of the new chain is performed after undergoing all moves dictated by the component kernels. For a move from θ to ϕ in a single iteration of a cycle chain, all possible moves to an intermediate state ψ_c through p_c, $c = 1, ..., r - 1$, finally leading to $\psi_r = \phi$ through p_r must be considered. Defining the initial state $\psi_0 = \theta$ gives

$$p(\theta, \phi) = \int \cdots \int \prod_{c=1}^{r} p_c(\psi_{c-1}, \psi_c) d\psi_1 ... d\psi_{r-1}$$

which generalizes results from section 4.6. Case (b) above is the special case of a cycle with $r = d$ and each transition p_m moves only the mth component of θ, $m = 1, ..., d$.

Many of the properties of the transitions p_c are passed on to the cycle transition p. First of all, the cycle kernel p defines a transition kernel of a Markov chain with equilibrium distribution π. Unlike mixture kernels, irreducibility and aperiodicity of one of the component transition kernels is not in general sufficient for irreducibility and aperiodicity of the cycle kernel. If all the component kernels are irreducible and aperiodic then the cycle kernel is irreducible and aperiodic (Tierney, 1994).

These forms can be integrated in to the Metropolis-Hastings algorithm. Each of the transition kernels p_i above may be given by a proposal kernel q_i and an acceptance probability α_i. Consider now the cycle scheme with componentwise transitions, namely, the component transition kernel $q_i(\theta, \phi)$ proposes a move of the ith component of θ, $i = 1, ..., r$. From Chapter 2, $\pi(\theta) = \pi_i(\theta_i)\pi(\theta_{-i})$ where π_i is the full conditional density of θ_i. The move determined by q_i only changes θ_i, so $\theta_{-i} = \phi_{-i}$ and $\pi(\phi)/\pi(\theta) = \pi_i(\phi_i)/\pi_i(\theta_i)$. Note that as far as the transition p_i is concerned, the other components of θ remain fixed and are not affected. Thus, it defines a reducible Markov chain. The proposal transition q_i may then be written in the form $q_i(\theta_i, \phi_i)$ even though it may well depend on the value of θ_{-i}. Consequently, the acceptance probability may also be written as

$$\alpha_i(\theta_i, \phi_i) = \min\left\{1, \frac{\pi_i(\phi_i)q_i(\phi_i, \theta_i)}{\pi_i(\theta_i)q_i(\theta_i, \phi_i)}\right\} \tag{6.5}$$

Note that each of the component transition kernels above defines a reversible chain with equilibrium distribution $\pi_i(\theta_i)$, $i = 1, ..., d$. Namely, each component transition kernel satisfies the equation

$$\pi(\phi_i|\theta_{-i}) = \int \pi(\theta_i|\theta_{-i})p_i(\theta_i, \phi_i)d\theta_i$$

Considering only two components, a move from $\theta = (\theta_1, \theta_2)$ to $\phi = (\phi_1, \phi_2)$ is formed after moves for the two components are operated according to their respective kernels. If π is a stationary distribution for this cycle kernel, it must satisfy

$$\pi(\phi) = \int \int \pi(\theta)p_1(\theta_1, \phi_1)p_2(\theta_2, \phi_2)d\theta \tag{6.6}$$

The right hand side of (6.6) can be rewritten as

$$\int \int \pi(\theta_1|\theta_2)\pi(\theta_2)p_1(\theta_1, \phi_1)p_2(\theta_2, \phi_2)d\theta_1 d\theta_2$$

$$= \int \pi(\theta_2)\left[\int \pi(\theta_1|\theta_2)p_1(\theta_1, \phi_1)d\theta_1\right]p_2(\theta_2, \phi_2)d\theta_2$$

$$= \int \pi(\theta_2)\pi(\phi_1|\theta_2)p_2(\theta_2,\phi_2)d\theta_2$$

$$= \int \pi(\phi_1)\pi(\theta_2|\phi_1)p_2(\theta_2,\phi_2)d\theta_2$$

$$= \pi(\phi_1)\int \pi(\theta_2|\phi_1)p_2(\theta_2,\phi_2)d\theta_2$$

$$= \pi(\phi_1)\pi(\phi_2|\phi_1)$$

$$= \pi(\phi),$$

confirming the validity of Equation (6.6). The second and fourth equalities above follow from the stationarity of the conditional densities and the others follow from basic probability and integration operations. The result can be extended to any number of components using induction on the same argument. The derivation above was based on absolutely continuous transition kernels for notational simplicity. All results remain valid with minor technical changes for the mixed kernels of the Metropolis-Hastings algorithm. It can also be shown that despite the reducibility of the component transition kernels, the cycle kernel is irreducible and aperiodic (Tierney, 1994). So, the limiting distribution π is unique.

Once that is done, a new, componentwise version of the Metropolis-Hastings algorothm is given by:

1. Initialize the iteration counter $j = 1$ and set the initial value of the chain $\theta^{(0)}$.

2. Initialize the component counter $i = 1$.

3. Move the ith component of the vector of states of the chain to a new value ϕ_i generated from the density $q_i(\theta_i^{(j-1)}, \phi_i)$.

4. Calculate the acceptance probability of the move $\alpha_i(\theta_i^{(j-1)}, \phi_i)$ given by (6.5). If the move is accepted, $\theta_i^{(j)} = \phi_i$. If the move is not accepted, $\theta_i^{(j)} = \theta_i^{(j-1)}$ and the chain does not move.

5. Change the counter from i to $i + 1$ and return to step 3 until $i = d$. When $i = d$, go to step 6.

6. Change the counter from j to $j+1$ and return to step 2 until convergence is reached.

In fact, this was the form of the algorithm originally proposed by Metropolis et al. (1953). The positions of the molecules were modified one by one according to a symmetric transition with uniform bivariate distribution centered at the previous position of the molecule. The algorithm was introduced in this book with a single global transition to unify the presentation with the work of Hastings (1970).

Example 6.1 Bennett, Racine-Poon and Wakefield (1996) compare many

MCMC schemes in the context of longitudinal data studies with a non-linear mean structure. The response y_{ij} of individual i at time t_{ij} is explained by the non-linear regression model

$$y_{ij} = f(\psi_i, t_{ij}) + \epsilon_{ij}$$

where $\epsilon_{ij} \sim N(0, \sigma^2)$ and ψ_i is the non-linear regression coefficient for individual i, $j = 1, ..., n_i$ and $i = 1, ..., m$. The common structure relating the individuals is given through a hierarchical model where the ψ_i are assumed to be a sample from a $N(\mu, W)$ distribution. The model is completed with a second level where independent prior distributions are specified for the hyperparameters $\mu \sim N(b, B)$, $\sigma^2 \sim IG(n_0/2, n_0 S_0/2)$ and $W \sim IW(n_W/2, n_W S_W/2)$. It is common in pharmacology studies to specify that concentration levels of substances introduced in a system are described by the non-linear equation

$$f((\psi_1, \psi_2, \psi_3), t) = \psi_1 + \frac{\psi_2 t}{\psi_3 + t}.$$

This represents a curve starting at ψ_1 when $t = 0$ and advancing to the asymptotic value $\psi_1 + \psi_2$ when $t \to \infty$. Bennett, Racine-Poon and Wakefield (1996) show that the full conditional distributions of the hyperparameters μ, σ^2 and W are conjugate and easy to sample from but the same is not true for the regression coefficients $\psi_1, ..., \psi_m$. For these parameters, they compare sampling directly from the full conditionals with the rejection and ratio-of-uniform methods and indirectly with independence proposals based on a normal approximation to the likelihood and with normal random walk proposals as described in section 6.2. They found that Gibbs sampling with the rejection method achieves convergence faster but at the expense of many rejections per iteration and conclude that the Metropolis-Hastings schemes are easier to implement and more efficient in terms of computing time. Their findings seem to provide an empirical echo to the intuitive point that whenever the model produces awkward full conditional distributions, one should avoid Gibbs sampling in favour of other MCMC schemes (section 5.3.5).

Example 6.2 Analysis of spatially distributed data is an area where simulation methods using Markov chains have been heavily used. A recent review with references is given by Besag and Green (1993). The models start by considering a region with sites or pixels $S_1, ..., S_d$. For each site, a variable of interest y_i is observed with mean $h(\theta_i)$, for some function h. This variable can be the number of cases of a disease in a county, a radio-carbon count in an archaeological site or an intensity measurement such as colour or frequency in an image of a medical exam. The complete set of responses forms the observed scene or image. These applications account for both discrete, continuous, univariate and multivariate responses.

Depending on the problem, different models can be contemplated. Typically, it is assumed that there is an underlying pattern formed by the values of $\theta = (\theta_1, ..., \theta_d)'$ associated with the observed scene. This pattern is corrupted by some random observation mechanism. The main effort of the inference is to remove this observational noise and recover the underlying, unobserved scene.

Most models commonly assume observational independence conditional on the unobserved image θ, that is, $l(\theta) = f(y_1, ..., y_d | \theta_1, ..., \theta_d) = \prod_i f(y_i | \theta_i)$. The spatial structure is specified through the prior $p(\theta)$. Other effects may also intervene in the problem. These include the effect of explanatory variables to account for fixed variation and unstructured pixelwise random effects to account for unspecified data heterogeneity.

Concentrating on the spatial structure, Besag, York and Mollié (1991) suggest the specification of a prior in a pairwise difference form

$$p(\theta) \propto \exp\left\{ -\sum_{i<j} w_{ij}\phi(\theta_i - \theta_j) \right\}$$

They present many possibilities and interpretations for the weights w_{ij} and for the function ϕ. In particular, they consider $w_{ij} = I(S_j \in N_i)$ and $\phi(x) = x^2/2W$ where N_i defines a neighbourhood of S_i, $i = 1, ..., d$.

When observing cases of a disease in a given region, an appropriate observational model is $y_i \sim Poi(e^{\theta_i})$ that leads to the posterior distribution for θ and W

$$\pi(\theta, W) \propto \prod_{i=1}^{n} \exp\{\theta_i y_i - e^{\theta_i}\} W^{-\frac{n}{2}} \exp\left\{ -\frac{1}{2W}\sum_{i<j}(\theta_i - \theta_j)^2 \right\} p(W) \quad (6.7)$$

where $p(W)$ is the prior distribution of W. The full conditional distribution of θ_i is

$$\pi_i(\theta_i) \propto \exp\left\{ \theta_i y_i - e^{\theta_i} - \frac{n_i}{2W}(\theta_i - \bar{\theta}_i)^2 \right\} \quad (6.8)$$

where n_i is the number of elements in N_i and $\bar{\theta}_i = (1/n_i)\sum_{\{j:w_{ij}=1\}} \theta_j$. Besag, York and Mollié (1991) use the Gibbs sampler with the rejection method to sample from (6.8). Green (1991) suggested using instead proposals in the form $q_i(\theta_i, \cdot) = f_N(\cdot; (1-a)b + a\theta_i, (1-a^2)c)$, $i = 1, ..., d$. The constants a, b and c are chosen so as to make the test ratio as constant as possible thus increasing the chances of acceptance of the proposed value. This form of proposal falls into the category of autoregressive chains described in section 6.2. Note that $a = 1$ gives the random walk with variance c. The guidance provided for the choice of this variance in the above section may be used here for arbitrary values of a.

The above description of the algorithm concentrated on the scheme where components are updated one by one in the order they are given in the parameter vector θ. The mixture and cycle schemes show that it is also possible to have many other orderings of the components. This also includes schemes where some of the components are systematically updated more often than other components. This may be because the less frequently updated components are more difficult to generate or because they are able to move more freely across the parameter space.

Another possibility worth exploring is to consider mixtures (or cycles) of different transition kernels. These typically apply to different components of θ but nothing in the theory prevents the use of conceptually different sampling schemes to update the same component or group of components in a single Markov chain. This does not generally guarantee faster convergence but Gelfand and Carlin (1995) provide a nice example where considerable improvement in convergence is achieved by mixing different sampling schemes in a single chain.

The different forms of proposal kernels described in the previous section are all relevant here. They may be used to construct each of the proposals q_i considered here. The same can be said about the adequacy of different forms for the situation under study. For some of the components of θ it may be more natural to use independence chains based on the prior distribution if there is enough information available for that component. For other components, the likelihood may suggest more appropriate proposals (section 6.5). Failing that, remaining components may be updated by a simple random walk proposal.

6.4.2 Metropolis within Gibbs

Transitions are based on the full conditional distributions of the components of θ in the basic componentwise scheme above. The more similar the proposal q_i and the density π, or equivalently the full conditional π_i, are the closer to 1 will the acceptance probabilities (6.5) be. This does not necessarily ensure fast convergence but may imply substantial computational savings.

In the basic Metropolis algorithm, it was typically not possible to find a sampling kernel q that could approximate π without error. Were that possible, direct sampling from π would be available. Here, the situation is different. It is possible that π has a complicated form that prevents direct sampling from it but (some of) its full conditional distributions π_i can be directly sampled from. In this case, a convenient choice of proposal is to take $q_i = \pi_i$. The proposed value for θ_i is drawn from its full conditional and accepted with probability 1. Hence, the computational burden of the calculation of (6.5) is avoided. If that can be done for all components of θ, their values will be all sampled from the corresponding full conditionals and accepted. This is the Gibbs sampler!

In general, the simplification obtained is very convenient but again no optimality results are available in that direction. Note that the Gibbs sampler only depends on the previous value of the other components, unlike the Metropolis-Hastings scheme that also depends on the previous value of the component being updated. Besag et al. (1995) point out that this extra ingredient may allow the construction of a more efficient sampling scheme.

Going back to direct sampling from full conditionals, in complex models it may be possible to establish conditional conjugacy for some but not all components of the parameter. In this case, only some components will be available for direct sampling. For the components that cannot be directly sampled, Muller (1991b) suggests that they are sampled from π_i by a Metropolis-Hasting (sub-)chain inside a Gibbs sampling cycle. So, these components are sampled from a proposal q_i with possible acceptance governed by probabilities (6.5). This process would last for T iterations until convergence and consequent generation from π_i, the limiting distribution of the sub-chain. This sampling scheme is widely known as Metropolis-within-Gibbs. Muller (1991b) uses the value $T = 5$ in applications but in fact the construction of the sub-chain is unnecessary as a single iteration is sufficient. Note that the case $T = 1$ reproduces the Metropolis-Hastings cycle scheme that visits each component once and was described in the previous sub-section. Nowadays, the version $T = 1$ is almost always used so that the Metropolis-within-Gibbs nomenclature is somewhat misleading. A more appropriate name to describe it could be Gibbs-within-Metropolis as some Gibbs updating schemes are used inside a componentwise Metropolis-Hasting scheme.

6.4.3 Blocking

So far, nothing has been said about the choice of blocks $\theta_1, ..., \theta_d$. The results obtained for the Gibbs sampler do not reproduce with the same ease here. In particular, it is not true that the larger the block the faster is the convergence. In fact, it is likely that blocking many parameters in a single group in highly multidimensional problems is more detrimental than beneficial.

Example 6.1 (continued). The parameters of the model are given by $\psi_1, ..., \psi_m$, μ, σ^2 and W. Note that each of these parameters is itself a collection of components with the exception of the scalar σ^2. The regression coefficients ψ_i and their mean μ are vector parameters and W is a matrix of parameters. Bennett, Racine-Poon and Wakefield (1996) used this blocking in all their sampling schemes except for the Gibbs sampler with the ratio-of-uniform method where they sample each scalar component of the ψ_i separately.

Consider the most common case of a random walk chain with proposal q_i for the component θ_i formed by a large number of components. As the number of components gets large, it becomes more likely to have components of θ_i falling well in the tails of π_i. This will produce a proposed value with very low density π_i and hence the test ratio will also be small. As a result, very few proposed values will be accepted and convergence will be very slow. Of course, decreasing the appropriate entries of the variance matrix of the proposal will inhibit such extreme components of θ_i being proposed but that requires tuning of each component of the variance matrix to avoid it. This task is far from easy and can become a time consuming exercise.

Similar reasoning can be applied to other forms of chains. There is no reason to believe that the proposal q_i will remain close to π_i as the dimension of θ_i increases. This leads to the same end result that increasing the dimension of a block increases the chances of rejection.

Again, there are no theoretical results here but the rule followed in applied work is to form small groups of correlated parameters that belong to the same context in the formulation of the model. A typical example is given in Example 6.1 above. The structure of the model made natural the choice of the blocks $\psi_1, ..., \psi_m, \mu, \sigma^2$ and W. With this structure, inference via Gibbs sampling or Metropolis-Hastings sampling may successfully proceed. Gathering all regression coefficients in a single block $\psi = (\psi_1, ..., \psi_n)'$ leads to the difficulties described above. Small blocks of ψs may also be formed and would alleviate these difficulties. In the context of this application, it is better to work with each vector of regression coefficient separately because they are all independent conditionally on μ and W. This point will be returned to in more detail below and in the next section.

The important point to make in practice is to block parameters whenever it is possible and needed. It is possible to block whenever the acceptance rate does not fall to very small values, namely single digit percentages. Gelman, Roberts and Gilks (1996) provide some theoretical considerations for aiming at an optimal acceptance rate of around 24% for random walk chains. This rate should be taken as an indication and not as a rule. Efficient chains with higher acceptance rates are also possible.

Blocking of parameters is only needed to break correlations. Parameters that are conditionally independent given other parameters need not be blocked. Remember from Chapter 5 that blocking was used to define more appropriate directions of moves for the chain instead of the unrelated moves along the axes. When components are (conditionally) independent, however, moves will already be made along the individual directions and no improvement is made by changing these directions.

6.4.4 Reparametrization

The same comments made about choice of parametrization in Chapter 5 for the Gibbs sampler are valid here. Good parametrizations are still useful in improving mixing of the chain and accelerating convergence. The apparent freedom of choice of proposals is not unrestricted. Very large displacements are very likely to be rejected as discussed above. Only moves that are compatible with the structure of the model and the form of the posterior distribution will be accepted. So, models with highly correlated parameters will only allow very small moves.

This is the pattern already observed for the Gibbs sampler. For the Metropolis-Hastings algorithm, larger moves may even be proposed but they will very likely fall in the tails of the posterior and will force a small value for the test ratio. The only possible acceptance of these large moves is for the case when they are directed according to the correlation structure of the posterior. In high-dimensional problems, the identification of these directions may be very difficult or time-consuming. So, the large moves required for fast convergence will lead to rejected points. If an efficient parametrization can be found, then these large moves will no longer be rejected and fast convergence may be achieved. The application to dynamic models below illustrates this point. Another successful use of reparametrization is the case of centering in generalized linear mixed models as discussed by Gelfand, Sahu and Carlin (1996). They show that centering in generalized models reduces correlation and improves convergence just as in normal linear mixed models (section 5.4.3).

6.5 Applications

This section gives some details of applications of the Metropolis-Hastings methodology for models commonly used in practice. With only a few exceptions, as soon as the model gets away from linearity and normality it loses conditional conjugacy. Chapter 5 showed that conjugacy is a very useful property in conjunction with the Gibbs sampler. It becomes difficult to sample from some of the full conditional distributions without them.

One possibility is to employ some of the resampling techniques, such as the rejection method. A very attractive alternative in terms of simplicity is the use of a Metropolis-Hastings sampling scheme for these awkward sampling distributions. Example 6.1 illustrates this point in the context of non-linearity. The applications below detail uses of the algorithm in the context of non-normal models.

6.5.1 Generalized linear mixed models

Models with random effects have been described in the previous chapter. When they also contain fixed effects, they are are called mixed models. A general but not unique form for generalized linear mixed models is given by

$$
\begin{aligned}
f(y_i|\theta_i) &= a(y_i)\exp\{y_i\theta_i + b(\theta_i)\} \text{ with } E(y_i|\theta_i) = -b'(\theta_i) = \mu_i \\
g(\mu_i) &= x_i'\beta + z_i'\gamma_i \quad, \quad i = 1,...,n
\end{aligned} \tag{6.9}
$$

where the link function g is differentiable, β is the d-dimensional vector of regression coefficients and γ_i is the r-dimensional random effect associated with observation y_i, $i = 1,...,n$. Associated with these parameters, there are d-dimensional and r-dimensional vectors of covariates x_i and z_i that explain the fixed and random variations in the levels of the observations y_i, $i = 1,...,n$. The model is completed with independent prior distributions $\beta \sim N(a,R)$, $\gamma_i|W \sim N(0,W)$, $i = 1,...,n$, and $W \sim IW(n/2,nS/2)$. Note that if $x_i = z_i$, $i = 1,...,n$, the model becomes a special case of the generalized linear hierarchical model (2.17) with $X_1 = \text{diag}(x_1',...,x_n')$, $\beta_1 = (\beta_{11}',...,\beta_{1n}')'$, $\beta_{1i} = \beta + \gamma_i$, $i = 1,...,n$, $X_2 = I_n$, $C = W$ and $\beta_2 = \beta$.

The natural division of the parameters in blocks is given by $\beta, \gamma_1,...,\gamma_n$ and W. Alternative divisions include the block $(\gamma_1,...,\gamma_n)$ or even the block $(\beta,\gamma_1,...,\gamma_n)$. The former is unnecessary as the random effects are conditionally independent. The latter removes the problems that may be associated with correlation between random and fixed components. However, it is very likely to lead to very low acceptance rates due to its high dimension unless complicated search exercises for an adequate proposal are undertaken. Even then, the solution is likely to be specific to the problem being analysed and will not be adequate for other data sets.

The only block with full conditional distribution in conjugate form is W. The other blocks have full conditional densities

$$
\begin{aligned}
\pi_\beta(\beta) &\propto \exp\left\{\sum_{i=1}^n y_i\theta_i + b(\theta_i)\right\} f_N(\beta; a, R) \\
&\propto \exp\left\{-\frac{1}{2}(\beta - a)'R^{-1}(\beta - a) + \sum_{i=1}^n y_i\theta_i + b(\theta_i)\right\} \tag{6.10} \\
\pi_{\gamma_i}(\gamma_i) &\propto \exp\{y_i\theta_i + b(\theta_i)\}f_N(\gamma_i; 0, W) \\
&\propto \exp\left\{-\frac{1}{2}\gamma_i'W^{-1}\gamma_i + y_i\theta_i + b(\theta_i)\right\} \quad, \quad i = 1,...,n \tag{6.11}
\end{aligned}
$$

None of these distributions is easy to sample from. Zeger and Karim (1991) use rejection sampling with normal envelopes for each of these full conditional distributions. It is very difficult in this situation to tune the constant A to provide a proper envelope over all the parameter space without sacrificing efficiency. Clayton (1996) explores the possibility of sampling each of these parameters componentwise. He uses the adaptive rejection method as in most cases of interest the densities above are log-concave.

Other possibilities are likely to involve some use of Metropolis-Hastings methodology. They include sampling from the prior as a proposal in an independence chain, or even a random walk chain with variances given by prior variance or some measure of likelihood uncertainty as the inverse information matrix. As previously discussed, these forms require optimization of the tuning constant, which may be time consuming.

Gamerman (1997) uses proposals based on the IRLS algorithm (section 3.2.2). Consider the start of iteration j of the chain with previous values $\beta^{(j-1)}, \gamma_1^{(j-1)}, ..., \gamma_n^{(j-1)}$ and $W^{(j-1)}$ and assume that all blocks are updated at every iteration in the order above.

The first step is the construction of the proposal q_β for the block β conditional on all the other parameters being fixed at current values. A vector of *adjusted* observations $\tilde{y} = \tilde{y}(\beta^{(j-1)})$ with corresponding matrix of *adjusted* variances $\tilde{V} = \tilde{V}(\beta^{(j-1)})$ is formed according to the IRLS algorithm and (6.9). An *adjusted* regression model is then formed with

$$\tilde{y}_i \sim N(x_i'\beta + z_i\gamma_i^{(j-1)}, \tilde{V}), i = 1, ..., n$$

Combining with the prior $\beta \sim N(a, R)$ leads to an *adjusted* posterior distribution $\tilde{\pi}_\beta(\beta) = N(m^{(j)}, C^{(j)})$ where $m^{(j)} = C^{(j)}(R^{-1}a + X'\tilde{V}^{-1}\tilde{y}^*)$ and $C^{(j)} = (R^{-1} + X'\tilde{V}^{-1}X)^{-1}$. The vector of *readjusted* observations $\tilde{y}^*$ has components $\tilde{y}_i^* = \tilde{y}_i - z_i'\gamma_i^{(j-1)}$, $i = 1, ..., n$. This additional adjustment to the observations is caused by the known displacements $z_i\gamma_i^{(j-1)}$, $i = 1, ..., n$, as calculations are conditional on the values of the γ_i. In many cases, $\tilde{\pi}_\beta$ is a good approximation to π_β. Therefore, it is reasonable to take the proposal $q_\beta(\beta^{(j-1)}, \cdot) = \tilde{\pi}_\beta(\cdot)$. A proposed value β^* can be generated and $q_\beta(\beta^{(j-1)}, \beta^*)$ can be calculated. Note that $q_\beta(\beta^{(j-1)}, \cdot)$ depends on $\beta^{(j-1)}$ but in a very intricate way (through the adjusted observations) and this proposal does not fit into any of the categories described in section 6.3. To calculate the test ratio, the values of $q_\beta(\beta^*, \beta^{(j-1)})$ and $\pi_\beta(\beta^*)/\pi_\beta(\beta^{(j-1)})$ are required. The first one is obtained by repeating the above procedure with β^* replacing $\beta^{(j-1)}$. The second is obtained from (6.10). Depending on the acceptance stage, $\beta^{(j)}$ is taken as β^* or $\beta^{(j-1)}$.

A similar approach is used to construct proposals q_{γ_i}, $i = 1, ..., n$. The *adjusted* observations $\tilde{y}_i = \tilde{y}_i(\gamma_i^{(j-1)})$ with *adjusted* variances $\tilde{V}_i = \tilde{V}_i(\gamma_i^{(j-1)})$ form the regression model $\tilde{y}_i \sim N(x_i'\beta^{(j)} + z_i\gamma_i, \tilde{V})$. Combining with the prior $\gamma_i \sim N(0, W)$ leads to the *adjusted* posterior $\tilde{\pi}_{\gamma_i}(\gamma_i) = N(m_i^{(j)}, C_i^{(j)})$ where $m_i^{(j)} = C_i^{(j)} z_i \tilde{V}_i^{-1}\tilde{y}_i^*$ and $C_i^{(j)} = (W^{-1} + z_i\tilde{V}_i^{-1}z_i)^{-1}$. Again, the *readjusted* observation is $\tilde{y}_i^* = \tilde{y}_i - x_i'\beta^{(j)}$. Taking as a proposal $q_{\gamma_i}(\gamma_i^{(j-1)}, \cdot) = \tilde{\pi}_{\gamma_i}(\cdot)$, a new value γ_i^* is proposed and $q_{\gamma_i}(\gamma_i^{(j-1)}, \gamma_i^*)$ can be calculated.

Again, $q_{\gamma_i}(\gamma_i^*, \cdot)$ is obtained by repeating the above procedures with γ_i^* replacing $\gamma_i^{(j-1)}$. The value of $\pi_{\gamma_i}(\gamma_i^*)/\pi_{\gamma_i}(\gamma_i^{(j-1)})$ is obtained using (6.11). The test ratio can be calculated and depending on the acceptance stage, $\gamma_i^{(j)}$ is taken as γ_i^* or $\gamma_i^{(j-1)}$. The procedure is repeated for $i = 1, ..., n$.

Finally, the value of $W^{(j)}$ is drawn directly from the full conditional distribution of W or $\Phi = W^{-1}$ given by

$$\pi_\Phi(\Phi) \propto \prod_{i=1}^{n} f_N(\gamma_i; 0, \Phi^{-1}) \, f_W(\Phi; n_W/2, n_W S_W/2)$$

$$\propto |\Phi|^{n/2} \exp\left\{ -\frac{1}{2} \sum_{i=1}^{n} \gamma_i' \Phi \gamma_i \right\} |\Phi|^{\frac{n_W -(r+1)}{2}} \exp\left\{ -\frac{1}{2} \mathrm{tr}(n_W S_W \Phi) \right\}$$

$$\propto |\Phi|^{[n+n_W -(r+1)]/2} \exp\left\{ -\frac{1}{2} \mathrm{tr}\left[\left(n_W S_W + \sum_{i=1}^{n} \gamma_i \gamma_i' \right) \Phi \right] \right\}$$

This is the expression of the $W[(n + n_W)/2, (n_W S_W + \sum_{i=1}^{n} \gamma_i \gamma_i')/2]$ distribution. Techniques for generation of a value of Φ from the Wishart distribution above were described in Chapter 1. A generated value of $W^{(j)}$ is obtained by inversion of the matrix Φ.

An important point is that this solution incorporates the structure of the problem into the construction of the proposal transition. This ensures that the chain moves will have direction and magnitude governed by the model. High acceptance rates are obtained as a result without compromising the amplitude of the chain moves and coverage of the relevant regions of the parameter space. The price paid in this case is the amount of additional calculation required. This may be unnecessary for models with a simpler structure but provides a general framework for analysis of any generalized linear mixed model.

Example 6.3 Crowder (1978, Table 3) presents a data set with proportions of germinated seeds in $n = 21$ plates. The data set is influenced by the explanatory variables type of seed (s), root extract (r) and an interaction between these covariates. A larger variability than that explained by the binomial model was also noted by Crowder (1978). One possible model for this overdispersion is obtained with random effects. Breslow and Clayton (1993, section 6.1) proposed to model the germination probabilities p_i associated with the ith plate through the logistic relation

$$\mathrm{logit}\,(p_i) = x_i' \beta + \gamma_i$$

where $x_i' = (1, s_i, r_i, s_i r_i)$ and $\gamma_i \sim N(0, W)$, $i = 1, ..., n$, are the univariate random effects modelling the overdispersion. The model is completed with a non-informative prior distribution $p(\beta, W) \propto 1/W$. Observe that W here is scalar and therefore its full conditional is an $IG(n/2, \sum_{i=1}^{n} \gamma_i^2/2)$ distribution. Figure 6.1 presents the marginal histograms for the components of

β for an analysis using the Metropolis-Hastings algorithm above. Table 6.1 presents numerical summaries of the posterior distribution along the equivalent ones from the penalized quasi-likelihood (PQL) analysis of Breslow and Clayton (1993). The estimates are very similar but the uncertainty in the Bayesian inference is always larger. This point had previously been noted in the context of generalized linear models by Dellaportas and Smith (1993).

Table 6.1. Estimation summary for Crowder's seeds data

Parameter	PQL Estimate (SE)	MCMC Estimate (SE)
Intercept	-0.542 (0.190)	-0.543 (0.197)
Seed coef.	0.077 (0.308)	0.074 (0.332)
Extract coef.	1.339 (0.270)	1.313 (0.274)
Interaction coef.	-0.825 (0.430)	-0.755 (0.431)
$\sqrt{W}$	0.313 (0.121)	0.278 (0.167)

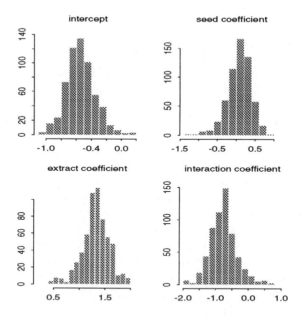

Figure 6.1. *Marginal histograms from the posterior distribution for the regression coefficients for the seeds data based on 600 successive draws from a single chain with a burn-in period of 600 iterations.*

Figure 6.2 presents a graphical summary of the inference for the random effects γ_i. Their posterior mean estimates behave as a sample from a normal distribution. So, they appear to confirm in the posterior the distributional form assumed for their prior. Some of the posterior correlations between the components of β were as large as 0.7 in absolute value but the correlation between β and W was low. This provides some support for the blocking scheme adopted, at least for this data set. The analysis presented was based on a single long chain. Similar results were however obtained by running multiple parallel chains.

This inference procedure can be easily extended to more elaborate forms of random effects. In many observation processes, data is obtained in groups. There may be random effects associated with groups (γ_i) and with individuals within groups (δ_{ij}). These may be described by the linear predictor

$$g(\mu_{ij}) = x'_{ij}\beta + z'_{ij}\gamma_i + t'_{ij}\delta_{ij} \ , j = 1, ..., n_i \ , i = 1, ..., m$$

An application of the above sampling techniques to this model is also presented in Gamerman (1997) and illustrated for a real data set. A very common special case is the so called Laird and Ware (1982) model where there are only random effects associated with groups.

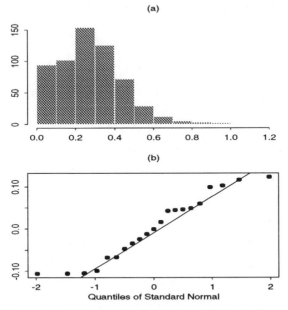

Figure 6.2. *Summary of inference for the random effects in the seeds data: (a) - marginal histogram of the posterior distribution of $\sqrt{W}$; (b) - QQ plot for normality of random effects. The posterior means for the random effects for each site were estimated from the sample and ordered for the construction of graph (b).*

6.5.2 Dynamic generalized linear models

Dynamic generalized linear models were introduced in Chapter 2. As previously seen, it is not possible to perform exact inference because the relevant marginal distributions cannot be obtained analytically. Assuming that the variances of the system disturbances are constant, the model parameters are given by the state parameters $\beta = (\beta_1, ..., \beta_n)'$ and the system variance $W = \Phi^{-1}$. The model is specified with the observation and system equations and completed with the independent prior distributions $\beta_1 \sim N(a, R)$ and $\Phi \sim W(n_W/2, n_W S_W/2)$. The option to work with the precision matrix instead of the variance matrix is made again. The posterior distribution is given by

$$\pi(\beta, \Phi) \propto \prod_{t=1}^{n} f(y_t|\beta_t) \prod_{i=2}^{n} p(\beta_t|\beta_{t-1}, \Phi)\, p(\beta_1)p(\Phi)$$

The full conditional distributions are given by:
a) for block β

$$\pi_\beta(\beta) \quad \propto \quad \prod_{t=1}^{n} f(y_t|\beta_t) \prod_{t=2}^{n} p(\beta_t|\beta_{t-1}, \Phi)\, p(\beta_1)$$

$$\propto \quad \exp\left\{ \sum_{t=1}^{n}[y_t\theta_t + b(\theta_t)] - \frac{1}{2}\sum_{t=1}^{n}(\beta_t - G_t\beta_{t-1})'\Phi(\beta_t - G_t\beta_{t-1}) \right\}$$

b) for block β_t, $t = 2, ..., n-1$

$$\pi_t(\beta_t) \quad \propto \quad f(y_t|\beta_t)\, p(\beta_t|\beta_{t-1}, \Phi)p(\beta_{t+1}|\beta_t, \Phi)$$

$$\propto \quad \exp\left\{y_t\theta_t + b(\theta_t)\right\} \exp\left\{ -\frac{1}{2}\left[(\beta_t - G_t\beta_{t-1})'\Phi(\beta_t - G_t\beta_{t-1}) \right.\right.$$
$$+ \quad (\beta_{t+1} - G_{t+1}\beta_t)'\Phi(\beta_{t+1} - G_{t+1}\beta_t) \left.\left.\right]\right\}$$

Similar results follow for blocks β_1 and β_n.
c) for block Φ

$$\pi_\Phi(\Phi) \quad \propto \quad \prod_{t=2}^{n} p(\beta_t|\beta_{t-1}, \Phi)\, p(\Phi)$$

$$\propto \quad \prod_{t=2}^{n} |\Phi|^{1/2} \exp\left\{ -\frac{1}{2}\mathrm{tr}[(\beta_t - G_t\beta_{t-1})(\beta_t - G_t\beta_{t-1})'\Phi] \right\}$$

$$\times \quad |\Phi|^{[n_W - (p+1)]/2} \exp\left\{ -\frac{1}{2}\mathrm{tr}(n_W S_W \Phi) \right\}$$

$$\propto \quad |\Phi|^{[n_W^* - (d+1)]/2} \exp\left\{ -\frac{1}{2}\mathrm{tr}\left[(n_W^* S_W^*)\, \Phi\right] \right\} \tag{6.12}$$

that is the density of the $W(n_W^*/2, n_W^* S_W^*/2)$ distribution with $n_W^* = n_W + n - 1$ and $n_W^* S_W^* = n_W S_W + \sum_{t=2}^{n}(\beta_t - G_t\beta_{t-1})(\beta_t - G_t\beta_{t-1})'$.

The results above show that the full conditional distributions of β and β_t do not belong to any known class of distributions but the full conditional of Φ is a known distribution from which samples can be drawn. As a result, the state parameters cannot be sampled directly. Again, rejection sampling can be used but the same problem of ensuring proper envelopes appears. The natural alternative seems to be the use of a Metropolis-Hastings scheme for this block. Therefore, transition kernels for the block β or for the blocks β_t, $t = 1, ..., n$, are required.

Again, many possibilities are available for the construction of the proposed kernels. Knorr-Held (1997) suggests the use of independence chains with prior proposals. He argues that the fast computing time at each iteration partially offsets the slow convergence due to the high correlation between state parameters. Shephard and Pitt (1997) use independence chains with proposals based on both prior and a normal approximation to the likelihood. They also block the state parameters β_t in random blocks to speed convergence. Proposal kernels may again be constructed with the IRLS algorithm used for evaluation of the posterior mode.

Singh and Roberts (1982) and Fahrmeir and Wagenpfeil (1997) extended to the dynamic setting the method of mode evaluation described in section 3.2.2 for static regression. They showed that iterating the posterior mode of β in the *adjusted* normal dynamic linear model given by (2.18)-(2.19) with *adjusted* observations $\tilde{y}_t$ and respective *adjusted* observational variances $\hat{V}_t$ leads to the posterior mode of β. The expressions of $\tilde{y}_t$ and $\hat{V}_t$ were given above.

The IRLS algorithm provides the *adjusted* full conditional distribution $\tilde{\pi}_\beta$ for block β given by (2.24). From this distribution, *adjusted* full conditional distributions for subsets of β may also be obtained. In particular, the *adjusted* distributions $\tilde{\pi}_t$ for blocks β_t, $t = 1, ..., n$, are given by (2.23). These distributions may be used as proposal kernels. They define a Markovian process as $\tilde{y}$ and $\tilde{V}$ depend on the value of $\beta^{(j-1)}$. Note that these are all multivariate normal distributions and therefore it is simple to draw from them. The draws are proposed and they may either be accepted or rejected depending on the acceptance probabilities. The complete sampling scheme for β and Φ is given by:

1. Initialize the iteration counter of the chain $j = 1$ and set initial values $\beta^{(0)} = (\beta_1^{(0)}, ..., \beta_n^{(n)})'$ and $W^{(0)}$.

2. Draw β^* from the density $\tilde{\pi}_\beta(\beta)$.

3. Calculate the acceptance probability $\alpha(\beta^{(j-1)}, \beta^*)$ of the move given by (6.5) with $q(\beta^{(j-1)}, \beta^*) = \tilde{\pi}_\beta(\beta^*)$. If the move is accepted, $\beta^{(j)} = \beta^*$. If the move is not accepted, $\beta^{(j)} = \beta^{(j-1)}$ and the chain does not move.

4. Draw Φ from its full conditional distribution (6.12).

5. Move counter from j to $j + 1$ and return to step 2 until convergence.

When updating with blocks $\beta_1, ..., \beta_n$ separately, steps 2 and 3 above are replaced by:

2'a. Initialize the component counter $t = 1$.

2'b. Draw β_t^* from the density $\tilde{\pi}_t(\beta_t)$.

2'c. Calculate the acceptance probability $\alpha_t(\beta_t^{(j-1)}, \beta_t^*)$ of the move given by (6.5) with $q_t(\beta_t^{(j-1)}, \beta^*) = \tilde{\pi}_t(\beta_t^*)$. If the move is accepted, $\beta_t^{(j)} = \beta_t^*$. If the move is not accepted, $\beta_t^{(j)} = \beta_t^{(j-1)}$ and the chain does not move.

3'. Move the counter from t to $t + 1$ and return to step 2'b until $t = n$. When $t = n$, go to step 4.

For normal models, it was shown that the Gibbs sampler operated over the block β is superior to the one operated over the blocks β_t. Although it is reasonable to expect the same behaviour here, there are important differences. The block β is highly dimensional for time series of large or moderate size. The high correlation between its components forces its complete conditional to be concentrated in a small region of the parameter space. It is very unlikely that proposed values are in this region and therefore, they are likely to fall well into the tails of this distribution. Consequently, their acceptance probability will be very low and the chain virtually does not move as a result.

That does not happen to blocks β_t of much smaller dimension. The proposal gives very good approximations for the full conditional distributions and high acceptance rates result. These high rates are not artificially obtained through small chain moves. They are governed by the structure of the model through the IRLS algorithm. The remaining problem is the high correlation between the components of β, making the chain move slowly towards equilibrium. Similar problems were found by Knorr-Held (1997).

An alternative previously discussed is the reparametrization in terms of the system disturbances w_t. The advantage again is that despite the high prior correlation between the β_t, the disturbances are independent a priori. Again, the components are dealt with separately but as most of the correlation is removed, the scheme is expected to converge at much faster rates. The drawback of the approach is the amount of extra calculations required by the reparametrization. Details about the method may be found in Gamerman (1998). Another possibility suggested by Shephard and Pitt (1997) is to form blocks containing small collections of β_t. The groups are divided at random which seems to improve convergence.

Example 6.4 The data set given in Table 6.1 and Figure 6.3 concerns a study on advertising awareness (Migon and Harrison, 1985). Samples of $n_t = 66$ people were selected at random every week for an opinion poll and asked whether they remembered having seen a given advertising campaign on TV. A weekly cumulative measure of campaign expenditure was constructed and is also depicted in Figure 6.3. The model used for this problem was a dynamic logistic regression

$$y_t \sim bin(n_t, \pi_t), \quad \mu_t = n_t \pi_t$$
$$\text{logit}(\pi_t) = \beta_{1t} + \beta_{2t} x_t = (1, x_t)\beta_t$$
$$\beta_t = \beta_{t-1} + w_t, \quad w_t \sim N(0, W)$$

Table 6.2 Data from Example 6.4

t	x_t	y_t	t	x_t	y_t	t	x_t	y_t
1	490	29	31	66	05	61	501	09
2	450	20	32	60	15	62	454	05
3	406	21	33	54	07	63	483	11
4	365	20	34	48	10	64	522	13
5	331	*	35	43	10	65	559	09
6	315	*	36	39	15	66	519	09
7	376	*	37	35	07	67	467	11
8	441	*	38	32	09	68	420	08
9	506	22	39	50	13	69	378	08
10	502	32	40	116	11	70	340	12
11	544	27	41	196	11	71	306	09
12	489	29	42	268	15	72	276	07
13	440	27	43	325	10	73	248	06
14	396	23	44	367	13	74	232	08
15	357	25	45	386	23	75	201	09
16	321	25	46	397	21	76	181	05
17	289	15	47	413	*	77	163	10
18	260	20	48	423	*	78	146	09
19	234	14	49	420	*	79	132	04
20	211	15	50	490	10	80	119	03
21	190	17	51	539	15	81	107	12
22	171	15	52	581	19	82	96	13
23	154	09	53	603	23	83	86	06
24	138	14	54	580	15	84	78	08
25	124	11	55	524	11	85	70	05
26	112	13	56	499	08	86	63	05
27	100	05	57	552	15	87	57	07
28	91	17	58	597	07	88	51	03
29	82	11	59	611	19	89	46	05
30	73	14	60	557	14	90	41	05

* - missing value.

The main features of this particular data set are a campaign change before week 41 and a few missing points for weeks during which the poll was not made.

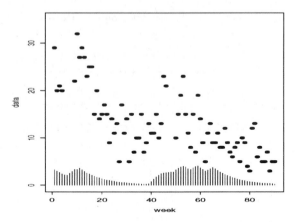

Figure 6.3. *Data on advertising awareness. The dots represent the weekly percentage of people that recalled having watched the advertising campaign of a given product on TV. The vertical bars represent a weekly cumulative measure of advertising expenditure.*

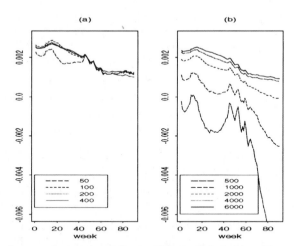

Figure 6.4. *Average trajectory of β_{2t} in 500 parallel chains with number of iterations for sampling from: (a) - system disturbances; (b) - state parameters.*

The average trajectory of the expenditure coefficient β_{2t} is plotted in Figure 6.4 for the sampling schemes based on the state parameters β_t and the system disturbances w_t. Convergence is clearly faster when sampling the disturbances, as expected. The drawback of higher computational cost is only serious for time series of very long size. The conceptual advantages seem stronger and should prevail in the choice of the method. Estimates for expenditure coefficients are plotted along with confidence limits in Figure 6.5. Two distinct levels are observed with a clear reduction after week 41, showing that the first campaign was more effective in terms of awareness. An increase in the uncertainty levels can be observed for the last weeks of the second campaign.

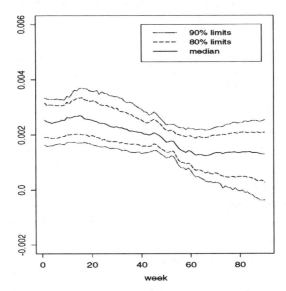

Figure 6.5. *Point estimate with confidence limits of the expenditure coefficient* β_{2t}.

6.6 Exercises

1. Prove that the transition kernel p of the Metropolis-Hastings algorithms satisfies the detailed balance equation (6.1) and hence has stationary distribution π.

2. The algorithm proposed by Barker (1965) set

$$\alpha(\theta, \phi) = \frac{\pi(\phi)}{\pi(\theta) + \pi(\phi)}$$

   Show that this algorithm produces a reversible chain and has stationary distribution π, if q is symmetric.

3. Certify yourself that large moves tend to be rejected and small moves are

very slow to converge by considering sampling from a N(0,1) distribution. Consider normal proposal transitions with variances ranging from 0.01 to 100. Compare also the independence chains with an arbitrary, fixed mean with the random walk chain with proposal means given by the previous value of the chain.

4. Certify yourself by graphical and/or analytical terms that if $q(\theta, \cdot)$ and $\pi(\cdot)$ are continuous, the acceptance probability will be close to 1 when moves proposed by q make the chain move slowly.

5. Consider random walk chains. Show that if the distribution f_w of the disturbances w_j is symmetric around 0, the chain is symmetric. Show also that the Metropolis algorithm described in section 6.1 is an example of a random walk chain and specify its distribution f_w.

6. Consider sampling π by the rejection method with an *envelope* density q for which complete blanketing is not assured and put it into the context of Metropolis-Hastings.

   (a) Defining the blanketing region $C = \{\theta | \pi(\theta) < Aq(\theta)\}$, obtain that

   $$\alpha(\theta, \phi) = \left\{ \begin{array}{ll} 1 & , \theta \in C \\ Aq(\theta)/\pi(\theta) & , \theta \notin C, \phi \in C \\ \min\{1, w(\phi)/w(\theta)\} & , \theta \notin C, \phi \notin C \end{array} \right.$$

   where $w = \pi/q$.

   (b) Discuss the computational advantages of the above scheme over independence chains with proposal q and over rejection sampling.

7. Consider r Markov chains with transition kernels p_i, $i = 1, ..., r$, with a common stationary distribution π. Show that the resulting kernel of the mixture of these transitions will also have stationary distribution π. Repeat the exercise for a cycle chain.

8. Consider a parameter vector $\theta = (\theta_1, ..., \theta_d)'$ with posterior density π and the componentwise Metropolis-Hastings algorithm with proposed kernels $q_i(\theta_i, \phi_i)$ and acceptance probabilities α_i given by (6.5), $i = 1, ..., d$.

   (a) Show that the component transition kernels formed define Markov chains with stationary distributions given by the full conditional distributions of θ_i, $i = 1, ..., d$.

   (b) Extend the results of section 6.4 to prove that π is a stationary distribution of the cycle kernel defined by the componentwise moves through all components of θ.

9. Consider the non-linear hierarchical model described in Example 6.1. Obtain the expressions of the full conditional densities for the hyperparameters μ, σ^2 and W and obtain the expressions of the proposed densities for the regression coefficients $\psi_1, ..., \psi_n$ that were discussed in the text.

10. Consider the spatial model described in Example 6.2. Show that:

    (a) the posterior distribution is given by (6.7);
    (b) the full conditional distribution of θ_i is given by (6.8);
    (c) the inverted Gamma distribution for W is conjugate conditionally on θ.

11. Describe in detail the sampling scheme for the dynamic generalized linear model for blocks $w_1, ..., w_n$ and Φ based on the IRLS algorithm. In particular, obtain the expressions of the proposal transition kernels and of the acceptance probabilities.

## CHAPTER 7

Further topics in MCMC

7.1 Introduction

The material presented in the previous chapters covers most of the relevant work on inferential procedures for a given model through Markov chain simulation techniques. Chapter 5 presented the Gibbs sampling technique and Chapter 6 presented the Metropolis-Hastings algorithm. It was assumed there that the adopted model was the true one or at least the most appropriate one throughout the presentation. Therefore, generation of a sample from *the* posterior distribution was all that was required. The techniques presented showed different ways of doing so.

In this final chapter, some points that lie beyond that basic framework will be discussed. Initially, the model will be put under scrutiny. Some techniques for evaluation of the model will be discussed. This evaluation may be divided into two complementary activities. The adequacy of a given model in the light of the observed data is made in section 7.2. A more encompassing treatment is presented in section 7.3 where different models are considered simultaneously. Depending on the cardinality and complexity of the set of models considered, recent methods based on Markov chains with jumps must be considered. Alterations in the structure of a given chain in order to speed up convergence are discussed in section 7.4. There are many ways of performing these changes, from alterations in the transition kernel to alterations in the target distribution. Other forms of change involve alteration in the generated sample. This chapter presents more advanced ideas that have not been used as much in practice. It has therefore a more speculative nature than the previous ones.

7.2 Model adequacy

Model evaluation necessarily depends on the verification of the fit of the data, or in other words, on the model likelihood. If the model likelihood is low then something must be redone, possibly some of the specifications that led to the model structure. This model likelihood was presented in Chapter 2 in Equation (2.1). It is usually referred to as predictive likelihood because it gives the expected value of the likelihood with respect to the prior. So, it is the mean prediction of the data probability before observing the data.

Suppose now that two models M_1 and M_0 are being contemplated as possible descriptions of the observed phenomenon with prior probabilities $p = Pr(M_0)$ and $1 - p = Pr(M_1)$. Given observations y, the posterior probabilities of the models satisfy

$$\frac{\pi(M_1)}{\pi(M_0)} = \frac{f(y|M_0)}{f(y|M_1)} \frac{1-p}{p}$$

where $\pi(M_i)$ is the posterior probability of model M_i and $f(y|M_i)$ is the likelihood of model M_i, $i = 0, 1$. Typically, the specification of model M_i involves parameters θ_i and the sampling distributions for observations y are different for the two models. In this case, the model likelihood $f(y|M_i)$ becomes a predictive likelihood and is given by

$$f(y|M_i) = \int f(y|\theta_i, M_i) p(\theta_i|M_i) d\theta_i \qquad (7.1)$$

where, unlike (2.1), Equation (7.1) takes explicitly into consideration the dependence on the model of all probability distributions.

The ratio $f(y|M_0)/f(y|M_1)$ of predictive likelihoods is called the Bayes factor. It provides the posterior odds ratio in favour of M_0 but for the prior odds ratio. When this is 1, indicating prior indifference (or equal chances), the posterior odds ratio coincides with the Bayes factor. So, the Bayes factor provides a relative measure for model evaluation. The calculations above are easily extended for the case of many possible models $M_1, ..., M_k$. A recent review of Bayes factors can be found in Kass and Raftery (1995).

An important aspect of model evaluation is the calculation of the predictive likelihood. Remember that it is the normalizing constant of the posterior distribution. Usually, its expression cannot be analytically obtained due to the complexity of the integrand in (7.1) and approximate methods must be used. Equation (3.3) provided an analytical approximation supported by asymptotic normal theory. In what follows, methods for approximate evaluation of (7.1) using simulation techniques will be presented. Different uses of these approximations for model evaluation will then be shown.

Although evaluation of a model presupposes the existence (or possibility) of other models, the calculations of this section will operate on a single model M at any one time. Therefore, the presence of the model in the conditioning part of probability statements will be supressed. Approaches that take into consideration different models simultaneously will be considered in the next section. In those cases, the model must be explicitly considered in the conditioning part of the distributions.

7.2.1 Estimates of the predictive likelihood

Equation (7.1) can be written as $f(y) = E[f(y|\theta)]$ where expectation is taken with respect to the prior distribution $p(\theta)$. The simplest estimate for

(7.1) consists of the basic Monte Carlo estimate

$$\hat{f}_1(y) = \frac{1}{n} \sum_{i=1}^{n} f(y|\theta_i)$$

where $\theta_1, ..., \theta_n$ is a sample from the prior $p(\theta)$. Raftery (1996) argues that this estimator does not work well in cases of disagreement between prior and likelihood, based also on applications by McCulloch and Rossi (1991). Almost all previous chapters contained some discussion of the difficulties associated with approximate inferences based on the prior, especially with sampling-based approaches. In the light of this information, it is not surprising to learn that $\hat{f}_1$ does not provide a sensible estimate. It averages likelihood values that are chosen according to the prior. In general, the likelihood is more concentrated than the prior and the majority of θ_i will be placed in low likelihood regions. Even for large values of n, this estimate will be influenced by a few sampled values, making it very unstable.

An alternative is to perform importance sampling with the aim of boosting sampled values in regions where the integrand is large. This approach is based on sampling from the importance density $g(\theta) = kg^*(\theta)$ where g^* is the unnormalized form of the density and k is a normalizing constant. Equation (7.1) can be rewritten as

$$f(y) = \int f(y|\theta) \frac{p(\theta)}{kg^*(\theta)} kg^*(\theta) d\theta$$

and therefore $f(y) = E_g[f(y|\theta)p(\theta)/kg^*(\theta)]$ where E_g denotes an expectation with respect to the importance distribution $g(\theta)$. This form motivates a new estimate

$$\hat{f}_2(y) = \frac{1}{n} \sum_{i=1}^{n} f(y|\theta_i) \frac{p(\theta_i)}{kg^*(\theta_i)}$$

where $\theta_1, ..., \theta_n$ is a sample from the importance density $g(\theta)$. In many cases, the value of k is not known and must be estimated. Noting that

$$k = \int kp(\theta) d\theta = \int \frac{p(\theta)}{g^*(\theta)} g(\theta) d\theta$$

leads to the estimator of k given by $\hat{k} = (1/n) \sum_{i=1}^{n} p(\theta_i)/g^*(\theta_i)$ where the θ_i are sampled from g. Replacing this estimate in the expression of $\hat{f}_2$ gives

$$\hat{f}_3(y) = \frac{\sum_{i=1}^{n} f(y|\theta_i)p(\theta_i)/g^*(\theta_i)}{\sum_{i=1}^{n} p(\theta_i)/g^*(\theta_i)}$$

These results can be applied in the Bayesian context by taking the posterior density π as the importance density. After a (Markov chain) simulation process, a sample $\theta_1, ..., \theta_n$ from $\pi(\theta) = kl(\theta)p(\theta)$ is available. These values can be used in $\hat{f}_3$ with $g^* = l \times p$. In this case, the estimator simplifies to

$$\hat{f}_4(y) = \left[\frac{1}{n} \sum_{i=1}^{n} \frac{1}{l(\theta_i)} \right]^{-1} \qquad (7.2)$$

This estimator is the harmonic mean of likelihood values originally proposed by Newton and Raftery (1994). Despite its consistency, this estimator is strongly affected by small likelihood values. Raftery (1996) relates this weakness to the occasional divergence of the variance of the denominator in (7.2). Newton and Raftery (1994) suggest using a mixture of prior and posterior as an importance density but that would mean sampling from both densities. The simplicity of $\hat{f}_4$ make it a very appealing estimator and its use is recommended provided the sample is large enough.

Another generalization of this estimator was obtained by Gelfand and Dey (1994). For any given density $g(\theta)$,

$$1 = \int g(\theta) d\theta = \int g(\theta) \frac{f(y)\pi(\theta)}{f(y|\theta)p(\theta)} d\theta$$

where, as before, p is the prior and π is the posterior density of θ. So,

$$f(y) = \left[\int \frac{g(\theta)}{f(y|\theta)p(\theta)} \pi(\theta) d\theta \right]^{-1}$$

Sampling $\theta_1, ..., \theta_n$ from π leads to the estimate

$$\hat{f}_5(y) = \left[\frac{1}{n} \sum_{i=1}^{n} \frac{g(\theta_i)}{f(y|\theta_i)p(\theta_i)} \right]^{-1} \qquad (7.3)$$

Even though the method is specified for any density g, appropriate choices are very important for a good practical implementation. Gelfand and Dey (1994) suggest using g as an importance density for the posterior and to take a normal or t distribution that approximates π with moments based on the sample of θ. Raftery (1996) presents a simple example where g was taken in product forms for each parameter component. The estimates obtained are highly inaccurate, showing that some skill is required in choosing g.

Another approximation to the model likelihood has previously been given in Chapter 3. The normal approximation to the posterior gives the estimate (3.3) for $f(y)$. This expression is based on the evaluation of the values of m, the posterior mode, and V, an asymptotic approximation for the posterior variance matrix. Sampling-based approximations for m and V can be constructed if a sample $\theta_1, ..., \theta_n$ from the posterior is available. The mode m can be estimated as the sample value for which π is largest, i.e., $\pi(m) = \max_i\{\pi(\theta_i)\}$. Similarly, estimates for the posterior variance matrix may be given in the case of an independent sample by

$$\hat{V} = \frac{1}{n} \sum_{i=1}^{n} (\theta_i - \bar{\theta})(\theta_i - \bar{\theta})' \text{ where } \bar{\theta} = \frac{1}{n} \sum_{i=1}^{n} \theta_i$$

Raftery (1996) discusses alternative calculations of the value of m when computation of π is expensive and of the value of V with the use of robust estimators.

Finally, a very simple estimate results from the fact that $f(y) = f(y|\theta) p(\theta)/\pi(\theta)$. Typically, $f(y|\theta)$ and $p(\theta)$ are easy to calculate but $\pi(\theta)$ is not. However, if a sample of π is available, some form of histogram smoothing can be applied to get an estimate of π. An alternative estimate of π was obtained by Chib (1995) in the context where full conditional densities are available in closed form, as in Gibbs sampling, for instance. Note first that

$$\pi(\theta_i|\theta_1, ..., \theta_{i-1}) = \int \cdots \int \pi(\theta_i|\theta_{-i})\pi(\theta_{i+1}, ..., \theta_d)d\theta_{i+1} \cdots d\theta_d, \ i = 2, ..., d,$$

suggests approximating $\pi(\theta_i|\theta_1, ..., \theta_{i-1})$ by $\hat{\pi}(\theta_i|\theta_1, ..., \theta_{i-1})$ given by

$$\hat{\pi}(\theta_i|\theta_1, ..., \theta_{i-1}) = \frac{1}{n} \sum_{j=1}^{n} \pi(\theta_i|\theta_1, ..., \theta_{i-1}, \theta_{i+1}^{(j)}, ..., \theta_d^{(j)})$$

where $(\theta_1^{(j)}, ..., \theta_d^{(j)})'$, $j = 1, ..., n$, is a sample from $\pi(\theta)$. As $\pi(\theta) = \pi(\theta_1) \prod_{i=1}^{d} \pi(\theta_i|\theta_1, ..., \theta_{i-1})$, an approximation $\hat{\pi}$ for the posterior is given by

$$\hat{\pi}(\theta) = \hat{\pi}(\theta_1) \prod_{i=1}^{d} \hat{\pi}(\theta_i|\theta_1, ..., \theta_{i-1})$$

Once an approximation $\hat{\pi}$ for π is available, it can be used to give another estimate

$$\hat{f}_6(y) = f(y|\theta)p(\theta)/\hat{\pi}(\theta)$$

Note that any value of θ can be used in the expression of $\hat{f}_6$ and if π could have been obtained without error, they would all provide the same estimate of $f(y)$. Obviously, θ should be chosen so that $\hat{\pi}$ has the smallest possible estimation error. This narrows the choice of θ to the central region of the posterior where π is likely to be estimated more accurately. Simple choices are the mode and the mean but any value in that region should be adequate.

Raftery (1996) reports comparisons between some of these estimates in a Ph. D. thesis by Rosenkranz (1992). The conclusions seem to favour the estimator based on normal approximations although with large samples the estimator $\hat{f}_4$ has behaved well. Other sampling-based estimators of $f(y)$ have been proposed by Raftery (1996), Gelfand and Dey (1994) and Meng and Wong (1993).

7.2.2 Uses of the predictive likelihood

When the prior distribution $p(\theta)$ is informative and proper, there is no problem in using $f(y)$ to evaluate different models. When the prior is improper, $f(y)$ cannot be calculated because the integral (7.1) diverges. In cases where the prior is proper but not very informative, the use of $f(y)$ as a tool for model evaluation should be made with care. So, the estimates obtained above for approximating $f(y)$ with weak prior information are likely to be unstable.

Even though the density (7.1) is the canonical form for model evaluation as prescribed by theory, it is not necessarily the only one. Different justifications have led to other approaches by several authors. Consider a sample $y = (y_1, ..., y_n)'$ and y_C denotes the subset of y containing the observations with indices in C. So, if $C = \{1, n\}$, $y_C = (y_1, y_n)'$ and if $C = \{1, ..., i - 1, i + 1, ..., n\}$, $y_C = y_{-i}$. Gelfand and Dey (1994) showed that many densities used for model evaluation can be written in the generic form

$$f(y_{S_1}|y_{S_2}) = \int f(y_{S_1}|\theta)p(\theta|y_{S_2})d\theta$$

$S_1 = \{1, ..., n\}$ and $S_2 = \phi$, the empty set, gives $f(y)$. If $S_1 = S_2 = \{1, ..., n\}$, then the density suggested by Aitkin (1991) for use in the posterior Bayes factor is obtained. The densities appearing in the definitions of the intrinsic Bayes factor of Berger and Pericchi (1996) and the fractional Bayes factor of O'Hagan (1995) also fit into this formulation. The main motivation for these alternative forms is their use with improper priors. An adequate choice of S_2 removes the impropriety of $p(\theta|y_{S_2})$ and therefore $f(y_{S_1}|y_{S_2})$ does not diverge and can be calculated.

Another density extensively used by Gelfand (1996) and Gelfand, Dey and Chang (1992) was the cross-validation predictive density $f(y_i|y_{-i})$ (Stone, 1974). Geisser and Eddy (1979) suggested the use of the product $\prod_{i=1}^{n} f(y_i|y_{-i})$ of these densities as a surrogate indicator of the value of the predictive likelihood $f(y)$ through the pseudo Bayes factor

$$\frac{\prod_{i=1}^{n} f(y_i|y_{-i}, M_0)}{\prod_{i=1}^{n} f(y_i|y_{-i}, M_1)}$$

that would approximate the Bayes factor.

Gelfand, Dey and Chang (1992) suggested many forms of use of the predictive densities through the expectations of functions $g(y_i)^*$ under $f(y_i|y_{-i})$. Among them are the prediction error $g_1(y_i) = y_i - y_{i,obs}$ where $y_{i,obs}$ is the observed value of y_i. The expectation of g_1 is $\gamma_{1i} = E(y_i|y_{-i}) - y_{i,obs}$. These values may be standardized to $\gamma'_{1i} = \gamma_{1i}/\sqrt{Var(y_i|y_{-i})}$. If the model is adequately fitting the data, the values of γ'_{1i} should be small.

---

* The reason why the functions g are considered instead of considering directly their expectations will be made clear below.

Under approximate normality, 95% of them should roughly lie between -2 and 2. The quantity $G_1 = \sum_{i=1}^{n} \gamma_{1i}'^2$ may be constructed and used as an indicator of model fit. The smaller its value, the better the fit of the model to the data.

Another useful function is $g_2(y_i) = I(y_i \leq y_{i,obs})$ with expectation $\gamma_{2i} = Pr(y_i \leq y_{i,obs}|y_{-i})$. Considering γ_{2i} as functions of $y_{i,obs}$ they are $U[0,1]$ distributed but they are not independent. So, the behaviour of a sample from a $U[0,1]$ distribution is expected. A large number of γ_{2i}s close to 0 or 1 indicates observations in the tails of their predictive densities, i.e., poor fit of the model, whereas a large number of γ_{2i}s close to $1/2$ indicates observations close to their corresponding predictive medians, i.e., good model fit. A possible summarization of the information from the γ_{2i}s is obtained with $G_2 = \sum_{i=1}^{n}(\gamma_{2i} - 0.5)^2$. Again, the smaller its value, the better the fit of the model to the data.

Similarly, the functions $g_3(y_i) = I(y_i \in \{y_i|f(y_i|y_{-i}) \leq f(y_{i,obs}|y_{-i})\}$ and $g_4(y_i) = I(y_i \in [y_{i,obs}-\epsilon, y_{i,obs}+\epsilon])/2\epsilon$ may be used. Their expectations are respectively given by $\gamma_{3i} = Pr(\{y_i|f(y_i|y_{-i}) \leq f(y_{i,obs}|y_{-i})\}|y_{-i})$ and $\gamma_{4i} = f(y_{i,obs}|y_{-i})$, when $\epsilon \to 0$. Again, as functions of $y_{i,obs}$, the γ_{3i}s are a sample from a $U[0,1]$ distribution and therefore can be summarized by $G_3 = \sum_{i=1}^{n}(\gamma_{3i} - 0.5)^2$. As for the γ_{4i}s, the natural summarizing quantity is their product, which is also present in the pseudo Bayes factor.

None of the functions of interest γ_{ji} can be obtained analytically for most models. However, sampling-based estimates can be obtained as they were written as expectations with respect to a distribution. Assuming the presence of a sample $y_{i1}, ..., y_{in}$ from $p(y_i|y_{-i})$, the γ_{ji} can be estimated by

$$\hat{\gamma}_{ji} = \frac{1}{n} \sum_{l=1}^{n} g_j(y_{il})$$

A sample from $p(y_i|y_{-i})$ can be obtained by noting that

$$p(y_i|y_{-i}) = \int p(y_i, \theta|y_{-i})d\theta$$
$$= \int p(y_i|\theta)p(\theta|y_{-i})d\theta$$

where the last equality follows from the usual assumption of conditional independence of the observations given θ. Therefore a draw from $p(y_i|y_{-i})$ is obtained by drawing θ_* from $p(\theta|y_{-i})$ and subsequently drawing y_i from $f(y_i|\theta_*)$ (section 1.3).

Only a sampling scheme for $p(\theta|y_{-i})$ remains to be described. Gelfand, Dey and Chang (1992) suggest the use of a resampling method (section 1.5) with some approximating density $q(\theta)$. A natural choice for q is the posterior density $\pi(\theta)$, for two reasons. For moderate to large samples, the exclusion of a single observation is not likely to significantly change the posterior. Therefore, $\pi(\theta)$ is likely to approximate $p(\theta|y_{-i})$ well. Also, a sample from π is already available from an inferential procedure for a given model.

Bayes' theorem with prior $p(\theta|y_{-i})$ and observation y_i gives $\pi(\theta) \propto f(y_i|\theta) p(\theta|y_{-i})$. Hence, following the construction of a sampling-importance resampling scheme in section 1.5 and assuming the presence of a sample $\theta_1, ..., \theta_n$ from π, the weights

$$w_j \propto \frac{\pi(\theta_j)}{p(\theta_j|y_{-i})} \propto \frac{1}{f(y_i|\theta_j)}$$

can be formed. These weights are normalized to add to 1 and used in the resampling scheme. The resulting sample has approximate distribution $p(\theta|y_{-i})$. A similar scheme can be devised for the rejection method with the disadvantage of finding a constant ensuring complete envelope. Gelfand (1996) also describes an alternative form to estimate the d_{ji} directly from a sample of π.

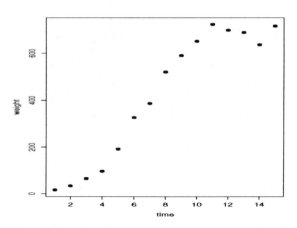

Figure 7.1. *Onion bulb data.*

Example 7.2 Figure 7.1 plots data presented by Ratkowski (1983) on the temporal evolution of the dry weight of onion bulbs. Gelfand, Dey and Chang (1992) consider 2 possible non-linear models in the form $y_t = f(\theta, t) + \epsilon_t$ for this data set: the logistic model $f(\theta, t) = \theta_1/(1 + \theta_2\theta_3^t)$ and the Gompertz model $f(\theta, t) = \theta_1 + e^{\theta_2\theta_3^t}$, where $\theta = (\theta_1, \theta_2, \theta_3)$. θ_1 represents the asymptote in both models but both θ_2 and θ_3 have different meanings under each model with $\theta_2 > 0$ and $0 < \theta_3 < 1$. The errors were assumed to be independent $N(0, \sigma^2)$ variables. The regression parameter θ was transformed to $\psi = (\psi_1, \psi_2, \psi_3)$ where $\psi_1 = \theta_1$, $\psi_2 = log\,\theta_2$ and $\psi_3 = log\,[\theta_3/(1 - \theta_3)]$ so that each of them varies over the real line. A non-informative prior $p(\psi, \sigma) \propto 1/\sigma$ was then assumed for each model, preventing evaluation of the model likelihoods.

Gelfand, Dey and Chang (1992) generated a sample of size 2000 of the 15 cross-validation predictive densities. They sampled from these densities using the weighted resampling technique. Rather than getting an initial sample from the posterior distribution, they approximated it by a t distribution and sampled from it. The parameters of the t distributions were set in accordance with a previous fit using the non-linear routines from the software SAS for each model. The individual values of the γ_{ji} were also provided for each model (Gelfand, Dey and Chang, 1992, Table 2). The summarizing quantities G_1, G_2 and G_3 are reproduced here in Table 7.1. The pseudo Bayes factor was approximated by 1.6863, indicating preference for the logistic model. This preference was confirmed by the other three criteria as shown in Table 7.1.

Table 7.1. Summarization of model evaluation

Criteria	Logistic	Gompertz
G_1	18.27	16.82
G_2	0.9219	1.3160
G_3	1.5657	1.9487

7.3 Model choice

Results from the previous section deal with evaluation of a given model. Comparisons of different models were also contemplated but always within the context where calculations were made separately for each model. In this section, a more formal treatment is given to the problem of choosing between models. This is done by indexing all models under consideration and treating this index as another parameter, to be treated jointly with all other model parameters.

Essentially, two alternative approaches are presented. The first was introduced by Carlin and Chib (1995) and considers all models in a formation, called here a supermodel. The Markov chain simulation scheme for this supermodel is presented below. The second approach presents sophisticated simulation techniques using Markov chain with jumps between the different models.

It will be assumed thorughout this section that y is observed and it can be described according to a model M_j with parameter θ_j of dimension d_j taking values in a parameter space $\Theta_j \subset R^{d_j}$, $j = 1, ..., J$. The value of J

could in principle be ∞ as, for instance, when considering countable classes of models.

A superparameter $\theta = (\theta_1', ..., \theta_J')'$ of dimension $d = \sum_{j=1}^{J} d_j$ taking values in a parameter space $\Theta = \cup_j \Theta_j \subset R^d$ and a quantity M assuming values in $\{1, ..., J\}$ can be defined. M serves the purpose of indicating a specific model. For instance, $M = j$ indicates that model M_j is being considered. It is implicitly assumed above that different models do not share any component of their respective parameters. This restrictive assumption is satisfied by many practical applications. Common components can also be included in this formulation with repetition of the components in all θ_j that contain them.

Example 7.2 (continued) The two models considered have an asymptote that is clearly common to both. Denoting the parameters of the logistic model by $(\theta_{1L}, \theta_{2L}, \theta_{3L}, \sigma_L^2)'$ and those of the Gompertz model by $(\theta_{1G}, \theta_{2G}, \theta_{3G}, \sigma_G^2)'$ gives $\theta_{1L} = \theta_{1G} = \theta_1$.

Assume for the moment that the posterior distribution $\pi(\theta, j)$, the joint distribution of the superparameter and the model indicator, is to be obtained. However, the main interest in inference is to obtain the posterior distributions of $\theta_j | M = j$, $j = 1, ..., J$, and M. These distributions respectively provide the posterior inference within each of the models and the posterior probabilities of the models. The joint posterior is useful when dealing with a component of the parameter, say ϕ, that is shared by all models. Inference about ϕ is based on its marginal posterior distribution $\pi(\phi)$. In any case, it can all be obtained from the joint density-probability function $\pi(\theta, j)$. The supermodel approach provides a sample from this more general, perhaps unnecessary posterior distribution whereas the approach with jumps only provides samples from $\theta_j | M = j$, $j = 1, ..., J$, and M. The presence of common parameters does not pose any problem here. Samples from them are naturally obtained from the corresponding samples for all models that include this shared component.

7.3.1 Markov chains for supermodels

The joint distribution of all random quantities is given by

$$p(y, \theta, j) = p(y|\theta, j)p(\theta|j)p_j$$

where j is the value of M and $p_j = Pr(M = j)$. Given that $M = j$, the distribution of y depends on θ only through θ_j, or mathematically,

$$p(y|\theta, j) = p(y|\theta_j, j)$$

Assume also that the θ_j are conditionally independent given the value of M. Hence,

$$p(\theta|j) = \prod_{i=1}^{J} p(\theta_i|j) \tag{7.4}$$

Note that the prior distribution $p(\theta_i|j)$, for $i \neq j$ does not make much sense. It specifies the distribution of the parameters of model i conditioned on the fact that this is not the true model. Carlin and Chib (1995) refer to these as pseudo prior or linking distributions. Due to the conditional independence (7.4), these priors do not interfere in the expressions of the marginal predictive densities for each model. Nevertheless, they are relevant for the construction of the chain and must be specified.

It follows from the above specification that

$$p(y, \theta, j) = p(y|\theta_j, j) \prod_{i=1}^{J} p(\theta_i|j) p_j$$

which is proportional to the joint posterior distribution of θ and M. A natural blocking is formed by grouping each model parameters and M. The full conditional distributions for $\theta_1, ..., \theta_J$ and M are obtained as follows:

- For block θ_j, $j = 1, ..., J$,

$$\pi_j(\theta_j) \propto \begin{cases} p(y|\theta_j, j) p(\theta_j|j) & \text{, for } M = j \\ p(\theta_j|i) & \text{, for } M = i \neq j \end{cases}$$

- For block M

$$\pi_M(j) = k^{-1} p(y|\theta_j, j) \prod_{i=1}^{J} p(\theta_i|j) p_j \, , \quad j = 1, ..., J$$

that is a discrete distribution with proportionality constant

$$k = \sum_{l=1}^{J} p(y|\theta_l, l) \prod_{i=1}^{J} p(\theta_i|l) p_l$$

M can always be sampled directly because it has a discrete distribution. Direct sampling from blocks θ_j will depend on the conjugacy structure for model $M = j$ and the form of the pseudo prior distributions. When direct sampling for some of the θ_js is not possible, Metropolis-Hastings steps described in Chapter 6 may be used.

The above scheme satisfies the conditions of a conventional Markov chain and therefore converges to the target distribution given by the posterior $\pi(\theta, j)$. Random draws from this distribution may be generated by iterative sampling from the full conditional distributions described above. The end result of this process is a sample of size n, say $\theta_1^{(l)}, ..., \theta_J^{(l)}, j^{(l)}$, $l = 1, ..., n$.

Comparison between models is based on the marginal posterior distribution of M, $\pi(j)$, $j = 1, .., J$. These probabilities are estimated by the proportion of values of M equal to j in the sample of size n.

Inference within each model is based on the conditional posterior distribution $\pi(\theta_j | j)$, $j = 1, ..., J$. Note that the sample available for θ_j is a sample from the marginal posterior $\pi(\theta_j)$. A sample from the conditional posterior is obtained by retaining the $\theta_j^{(l)}$, $l = 1, ..., n$, of the posterior sample for which the sampled value for M was j and discarding all other values of $\theta_j^{(l)}$ for which the sampled value for M was not j.

The pseudo prior distributions must be carefully chosen as they affect the rate of convergence of the chain. Note that as a modelling device they are meaningless. Therefore, Carlin and Chib (1995) support the view that total freedom may be given to the specification of these prior distributions. They may even include specifications using the data. After experimenting with a few choices, they recommend setting the pseudo prior distributions $p(\theta_j | i)$, for $i \neq j$ as close as possible to $\pi(\theta_j | j)$. They suggest the use of simple standard approximations based on univariate estimates obtained from pilot (model-specific) chains.

Another difficulty encountered by Carlin and Chib (1995) is connected to the prior model probabilities p_j. They observe that for some prior specifications, the chain does not seem to move between models and, as a result, large posterior probabilities are obtained for some of the models. To correct that, they set the prior probabilties in their examples to values that allow movement between the models. Although this may work in practice, it can force practitioners to specify probabilities they may not believe. Their example contained fairly vague prior distributions for models with different dimensions and Bayes factors are known to be very sensitive in these situations. So, this prior setting may need further justification to satisfy potential users.

Finally, this approach is not applicable to the case of a countable number of models under consideration. Hence, the amount of practical and theoretical difficulties of this approach suggests it should be used with care.

7.3.2 Markov chains with jumps

An alternative route for general construction of Markovian processes with a given stationary distribution is based on the specification of a transition kernel satisfying the detailed balance equation

$$\pi(\theta, j)p((\theta, j), (\phi, k)) = \pi(\phi, k)p((\phi, k), (\theta, j))$$

valid for all points where the move is allowed. This equation ensures reversibility of the chain. This is a sufficient condition and hence imposes more restrictions than necessary for convergence. Nevertheless, it provides a useful basis for specification of appropriate transition kernels. This route was used by Hastings (1970) and described in Chapter 6. It led to many possibilities for the kernel and is again used here in the more general context of a collection of models.

Once again, the transition will be constructed in two stages: a proposal transition and an acceptance probability, correcting the proposal to ensure balance. The main difference here is that many models are simultaneously being considered and therefore many qualitatively different moves are entertained. Green (1995) explored this idea by imposing detailed balance for all possible moves between models. Detailed balance would then be attained globally and convergence to $\pi(\theta, j)$ would result.

Consider for each possible jump move m between models, an arbitrary transition kernel $q_m((\theta, j), (\phi, k))$ and a yet to be specified acceptance probability $\alpha_m((\theta, j), (\phi, k))$. In fact, many different jump moves can be made from M_j to M_k but for simplicity it will be assumed here that each move m uniquely defines a model M_k. Many of these moves can be specified and for each of them

$$B_{jm} = \int_{\Theta} q_m((\theta, j), (\phi, k))d\phi$$

can be defined. B_{jm} gives the probability that the proposed move makes the chain jump from current model M_j to model M_k and may depend on θ. In practice, this is typically not the case and dependence on θ is dropped for notational simplicity. Note now that $q_m((\theta, j), (\cdot, k))/B_{jm}$ defines a proposal transition density for θ_k associated with jump move m.

Each proposed move takes the chain from model j and a corresponding value of θ_j to model k and a corresponding value of ϕ_k. Only values of the parameters associated with the models involved in the jump are concerned. If the move is accepted, only the kth component of θ is altered and $\phi_{-k} = \theta_{-k}$. Note also that k may be equal to j and, in this case, the move does not involve a jump between models.

The extension considered by Green (1995) also admits that the chain may not move at every iteration so that $\sum_m B_{jm} = B_j \leq 1$. Jump moves m are proposed according to probabilities B_{jm} and there is also a probability $1 - B_j$ that the transition does not propose any moves. Naturally, it is possible to have $B_j = 1$ and in this case a move will always be proposed.

Following (6.2), the transition kernel of the chain is given by

$$p((\theta, j), A) = \sum_m \int_A q_m((\theta, j), (\phi, k))\alpha_m((\theta, j), (\phi, k))d\phi +$$
$$I((\theta, j) \in A)s(\theta, j) \tag{7.5}$$

where $A \subset \Theta \times \{1, ..., J\}$ and

$$s(\theta, j) = 1 - \sum_m \int_{\Theta} q_m((\theta, j), (\phi, k))\alpha_m((\theta, j), (\phi, k))d\phi$$

$$= \sum_m \int_\Theta q_m((\theta, j), (\phi, k))[1 - \alpha_m((\theta, j), (\phi, k))]d\phi \quad (7.6)$$

$$+ \quad 1 - B_j \quad (7.7)$$

As for the Metropolis-Hastings algorithm presented in Chapter 6, the transition kernel (7.5) defines a mixed distribution for the next state of the chain. For points $(\phi, k) \neq (\theta, j)$, the distribution admits a combination of a density for ϕ_k and a probability function for k given by $q_m((\theta,j),\cdot)\alpha_m((\theta,j),\cdot)$ where m is the jump move taking the chain from model M_j to model M_k. More specifically, there is a probability B_{jm} that the proposed move takes the chain to model M_k. Given that the proposed move took the chain to model M_k, there is a density g describing the chances of moves to ϕ_k. Associated with the jump m to model M_k, the superparameter can be partitioned as $\theta = (\theta_k, \theta_{-k})$. The density g is given by

$$g(\phi_k) = q_m((\theta_k, \theta_{-k}, j), (\phi_k, \theta_{-k}, k))/B_{jm} \quad (7.8)$$

This density governs the proposed value for the parameter of model k. This value will then be accepted with probability $\alpha_m((\theta_k, \theta_{-k}, j), (\phi_k, \theta_{-k}, k))$. Note that both the proposal q_m and the acceptance probability α_m depend on θ only through θ_k; the remaining components are irrelevant once the jump defines model M_k for the next state of the chain.

For $(\phi, k) = (\theta, j)$, this distribution provides a positive probability of no moves given by $s(\theta, j)$. Equation (7.7) informs that this probability is due to two distinct events: either the move took the chain to a state that was not accepted or no move was proposed.

The imposition of general reversibility of the process leads to

$$\alpha_m((\theta, j), (\phi, k)) = \min\left\{1, \frac{\pi(\phi, k)q_m((\phi, k), (\theta, j))}{\pi(\theta, j)q_m((\theta, j), (\phi, k))}\right\} \quad (7.9)$$

In practical terms, the simulation of a sample from π using the Markov chain defined by transition (7.5) may be summarized as:

1. Initialize the iterations counter $l = 1$ and set as arbitrary an initial value $(\theta^{(0)}, j^{(0)})$.

2. Choose a jump move m accordingly with probabilities B_{jm}, thus defining the model M_k considered for move proposals. Hence, the proposed value for $j^{(l)}$ is k. If the move chosen was not to move, set $(\theta^{(l)}, j^{(l)}) = (\theta^{(l-1)}, j^{(l-1)})$ and go to step 5. This happens with probability $1 - B_j$.

3. Draw ϕ_k from the conditional density (7.8). Hence, the proposed new state for θ changes its kth component to ϕ_k while keeping $\phi_{-k} = \theta_{-k}^{(l-1)}$. Hence, all other components of θ are unchanged.

4. Calculate the acceptance probability $\alpha_m((\theta^{(l-1)}, j^{(l-1)}), (\phi, k))$ of the move given by (7.9). If the move is accepted, $(\theta^{(l)}, j^{(l)}) = (\phi, k)$. If the

move is not accepted, $(\theta^{(l)}, j^{(l)}) = (\theta^{(l-1)}, j^{(l-1)})$ and the chain does not move.

5. Change counter from l to $l+1$ and return to step 2 until convergence is reached.

Steps 2 and 4 are operated after generation of two independent uniform random quantities u_1 and u_2. The first one is used to determine the model to which the chain will propose a move. This choice is made according to the discrete distribution with probabilities B_{jm}. The second random quantity determines the acceptance probability of the proposed move as before: if $u_2 \leq \alpha_m$, the move is accepted and if $u_2 > \alpha_m$ the move is not allowed.

This process in fact generates a stream of values $\theta_{j^{(l)}}^{(l)}, j^{(l)}, l = 1, 2, ...$ At each iteration once the value of the model is drawn, only the parameter associated with this model is generated. Therefore, samples from $\theta_j | j$ are automatically provided by restricting attention to the chain values associated with value j for the model. Likewise, samples from M are provided by the marginal samples over all iterations. At each iteration, the chain can be complemented with the current values of the parameters for the other models. This forms a larger stream of values $\theta^{(l)}, j^{(l)}, l = 1, 2, ...$, as in the previous section. There is no harm, of course, in doing it even though this is in general irrelevant as most questions of interest are already answered by the smaller and variable dimension sequence.

In the case of J models, $J + 1$ moves can be specified: $J - 1$ jumps to the other models, one move within a model and one absence of move. As J may be large, this may imply too many alternative moves. In general, few moves are necessary. The important point is to ensure irreducibility of the chain and freedom of movement throughout the parameter space. Typically, this is achieved by only allowing moves between neighbouring models. In the case of a special meaning associated with the ordering $1, 2, ..., J$ of the models, Phillips and Smith (1996) and Green (1995) suggest considering only jumps to models $j - 1$ and $j + 1$ from model j. Markov chains of this kind were studied in Chapter 4 where they were referred to as birth and death processes. Note also that in the presence of a single model with moves always being proposed, the above structure reduces to the Metropolis-Hastings algorithm described in the previous chapter.

Example 7.3 Choosing between two competing models

Consider models M_1 and M_2 with parameters θ_1 and θ_2 of dimensions d_1 and d_2 satisfying $d_1 + n_1 = d_2$ where $n_1 > 0$. So, model M_2 has a parameter space n_1 dimensions smaller than model M_1. The reversibility condition requires an additional random quantity u of dimension n_1 in order to define a bijection $\theta_2 = g(\theta_1, u)$, thus allowing moves in both directions with the same probability. There are many other ways to ensure reversibility discussed by Green (1995) but this is the simplest one. The transition consists of incorporating into the ratio test the information about this bijection and about the distribution f of u that may depend on θ_1. The acceptance probability of the move from model 1 to model 2 is given by

$$\min\left\{1, \frac{\pi(\theta_2, 2)B_{21}}{\pi(\theta_1, 1)B_{12}q(u|\theta_1)}\left|\frac{\partial g(\theta_1, u)}{\partial(\theta_1, u)}\right|\right\} \qquad (7.10)$$

where B_{ij} is the probability that a move from model i to model j is proposed, $q(u|\theta_1)$ is the conditional proposal density used to generate the additional quantity u and $\frac{\partial g(\theta_1, u)}{\partial(\theta_1, u)}$ is the matrix of derivatives of the bijection $g(\theta_1, u)$. Likewise, the acceptance probability of the reverse move from model 2 to model 1 is given by

$$\min\left\{1, \frac{\pi(\theta_1, 1)B_{12}q(u|\theta_1)}{\pi(\theta_2, 2)B_{21}}\left|\frac{\partial(\theta_1, u)}{\partial\theta_2}\right|\right\}$$

Details of this derivation are given by Green (1995).

The extension to more than two models follows the same ideas with specifications of quantities u_{ij} of appropriate dimensions to ensure bijections between the parameters of models M_i and M_j. So, if θ_i (θ_j) is the parameter of model M_i (M_j) with dimension d_i (d_j) and $d_j - d_i = n_{ij} > 0$, a n_{ij}-dimensional random quantity u_{ij} is defined. Then, a deterministic bijection is created between (θ_i, u_{ij}) and θ_j. This simple device ensures that the moves between any two parameter spaces can be reversed. There is no theoretical restriction on the form of the proposal densities used to draw the u_{ij} and of the bijection. Once again, they should lead to moves that are not too small (to render slow convergence) nor too large (to render low acceptance probability). Green (1995) applies this idea to models with multiple change points, image segmentation and partition models. He also discusses some useful forms of bijections and proposals in the context of his applications (see also Example 7.4 below).

In many situations, there is a natural ordering of the component models, usually through the dimensions of their parameter spaces. So, simple chains based on irreducible birth and death processes can be used. This approach seems to be very promising for the treatment of problems with many possible models such as mixtures with an unknown number of components (Richardson and Green, 1997), variable selection in regression (Dennison, Mallick and Smith, 1996) and non-parametric intensity rate estimation (Green, 1995; Arjas and Heikkinen, 1997).

Example 7.4 Non-parametric intensity rate estimation

Consider the observation of a Poisson process over $[0, T]$ with unknown intensity rate $\lambda(t)$, $0 \le t \le T$. Each of countably many possible models M_k, $k = 0, 1, 2, ...$, is defined as having a piecewise constant intensity rate taking values λ_j at intervals $I_j = [t_j, t_{j+1})$, $j = 0, 1, ..., k$, with $t_0 = 0$ and $t_{k+1} =$

T. The parameter of M_k is $\theta_k = (\lambda(k), t(k))'$ where $\lambda(k) = (\lambda_0, \lambda_1, ..., \lambda_k)$ and $t(k) = (t_1, ..., t_k)$. The hierarchical prior used by Green (1995) assumed that, conditional on k, $t(k)$ consists of the even-numbered order statistics of a sample of size $2k + 1$ from the $U[0, T]$ distribution and $\lambda(k)$ is a sample from a $G(\alpha, \beta)$ distribution. The prior is completed with a second stage $k \sim Poi(\gamma)$ with probability function f_P. Arjas and Heikkinen (1997) replaced the conditional independence of the $\lambda(k)$ by a pairwise difference prior accounting for their spatial interaction as in Example 6.2.

Birth and death processes allow for four different types of moves in these circumstances:

1. The birth of a new step, thus creating a new interval and moving the chain from model M_k to model M_{k+1}. This move has probability p_k.

2. The death of an existing step, thus deleting an existing interval and moving the chain from model M_k to model M_{k-1}. This move has probability q_k.

3. The change of the intensity rate of a given interval. This move has probability r_{1k}.

4. The change of the length of a given interval. This move has probability r_{2k}.

Note that $p_k + q_k + r_{1k} + r_{2k} = 1$ ($q_0 = 0$) and that moves of type 3 and 4 retain model M_k. For these moves, assuming randomly chosen intervals for the change, evaluation of acceptance probabilities follows from standard theory and is left as an exercise.

A birth move changes model M_k with parameter θ_k to model M_{k+1} with parameter θ_{k+1}. Green (1995) suggests starting the move by choosing a new endpoint t^* uniformly over $[0, T]$ (and this point will lie in the interval $[t_j, t_{j+1})$, say) and choosing the corresponding new rates λ'_j and λ'_{j+1} according to $\lambda'_j / \lambda'_{j+1} = u/(1-u)$ where $u \sim U[0, 1]$ but preserving the geometric average so that $(t_{j+1} - t^*)log\lambda'_{j+1} + (t^* - t_j)log\lambda'_j = (t_{j+1} - t_j)log\lambda_j$. Note that only two new variables were needed to create the two new parameters. Reversibility automatically defines the new intensity rate of the interval formed by the merging of two adjacent intervals in the death move. The endpoint to be deleted is chosen uniformly at random and the merged intensity rate should also satisfy the same geometric averaging.

Following (7.10), the acceptance probability of a birth move is

$$\min\left\{1, \frac{\pi(\theta_{k+1}, k+1)}{\pi(\theta_k, k)} \frac{B_{k+1,k}}{B_{k,k+1}q(t^*, u)} \left|\frac{\partial\theta_{k+1}}{\partial(\theta_k, t^*, u)}\right|\right\}$$

Derivation of the expressions of each of the terms above and of the acceptance probability of a death move are left as exercises.

A similar treatment to the problem of inference about parameters in different models and model choice was given by Phillips and Smith (1996) based on jump diffusion processes (Grenander and Miller, 1994). In these processes, the chain moves within a model following a diffusion process, namely a Markovian process at continuous time. Sampling from this process requires a discretization of time and consequent approximation of the stochastic differential equation governing the process by a difference equation similar to the one used to define a system equation in dynamic models. A jump process is superimposed onto this diffusion to allow for moves between models. The jumps occur according to the marginal jump intensity. This intensity depends on the state (θ, j) of the chain and is obtained by the integration of the jump intensity $q((\theta, j), (\phi, k))$ over all possible jump points (ϕ, k). As before, this intensity q is constructed so as to ensure reversibility. Many possibilities for q based on Gibbs samplers and Metropolis-Hastings algorithms are presented by Phillips and Smith (1996) and illustrated in the context of identification of mixture components, object recognition, variable selection and identification of change points.

7.4 Convergence acceleration

The previous section dealing with techniques for improving the convergence of the chain basically considered them in the context of blocking parameters and reparametrization. There are many other suggestions for improving the convergence of the chain to the equilibrium distribution. Some of these techniques are described in this section. For presentation purposes they are divided in two large groups: alterations in the chain and alterations in the equilibrium distribution.

The chain may be altered by specification of alternative transition kernels with improved convergence properties or by manipulation of the draws generated from the chain. In both cases, the goal is the same: to make the chain get to the equilibrium faster.

7.4.1 Alterations to the chain

One of the greatest problems for the convergence of the chain is the fact that it moves the components along the directions determined by the components of θ. One alteration of this scheme is reparametrization, which essentially promotes a transformation of the parameter space and can be seen as a redefinition of the sampling axes.

Rather than seeking useful reparametrization that will then define more appropriate directions for sampling, new directions can be suggested directly. Schmeiser and Chen (1991) proposed a chain where the direction e of the moves is chosen at random from the unit vectors in R^d where d is the dimension of θ. Once a direction is chosen, the next state of the chain is chosen in that direction according to the probabilities given by π. The complete scheme is:

1. Initialize the iteration counter of the chain $j = 1$ and set an initial value $\theta^{(0)}$.

2. Choose a direction $e^{(j)}$ in R^d uniformly on $\{(x_1, ..., x_d) : \sum_{i=1}^{d} x_i^2 = 1\}$.

3. Choose a scalar $c^{(j)}$ generated from the density $g(c) \propto \pi(\theta^{(j-1)} + ce^{(j)}))$ and set $\theta^{(j)} = \theta^{(j-1)} + c^{(j)}e^{(j)}$.

4. Change counter from j to $j + 1$ and return to step 2 until convergence is reached.

The generation of e can be made univariately by drawing r_i from some distribution symmetric around 0 and taking $e_i = r_i / \sum_j r_j^2$, $i = 1, ..., d$. Once the direction e is chosen, the amount c of movement along e is chosen according to the posterior distribution. No matter how complicated is the form of the posterior, the generation of c is univariate.

Example 7.5 Assume that the posterior distribution of interest is a bivariate mixture of normals $wN(\mu_1, \Sigma) + (1 - w)N(\mu_2, \Sigma)$ where $0 < w < 1$ and

$$\mu_i = \begin{pmatrix} \mu_{i1} \\ \mu_{i2} \end{pmatrix}, i = 1, 2, \text{ and } \Sigma = \begin{pmatrix} \sigma_{11} & \sigma_{12} \\ \sigma_{21} & \sigma_{22} \end{pmatrix}.$$

At iteration j and given direction $e^{(j)} = (e_1, e_2)'$, the generating density for $c^{(j)}$ is

$$g(c) \propto w \exp\left\{-\frac{1}{2}(\theta^{(j-1)} + ce^{(j)} - \mu_1)'\Sigma^{-1}(\theta^{(j-1)} + ce^{(j)} - \mu_1)\right\}$$

$$+ (1 - w) \exp\left\{-\frac{1}{2}(\theta^{(j-1)} + ce^{(j)} - \mu_2)'\Sigma^{-1}(\theta^{(j-1)} + ce^{(j)} - \mu_2)\right\}$$

$$\propto w \exp\left\{-\frac{1}{2}(rc^2 - 2cs_1 + t_1)\right\} + (1 - w) \exp\left\{-\frac{1}{2}(rc^2 - 2cs_2 + t_2)\right\}$$

where $r = e^{(j)'}\Sigma^{-1}e^{(j)}$, $s_i = e^{(j)'}\Sigma^{-1}(\theta^{(j-1)} - \mu_1)$ and $t_i = (\theta^{(j-1)} - \mu_i)'\Sigma^{-1}(\theta^{(j-1)} - \mu_i)$, $i = 1, 2$. It is clear that g is the density of a mixture of normal distributions with means s_i/r, $i = 1, 2$, weights w_1 and $1 - w_1$ where $w_1 = we^{(s_1^2/2r)-(t_1/2)}/[we^{(s_1^2/2r)-(t_1/2)} + (1 - w)e^{(s_2^2/2r)-(t_2/2)}]$ and common variance r^{-1}. So, draws of c are easily obtained.

A generalization of this algorithm was proposed by Phillips and Smith (1993). Each iteration in their sampling scheme comprises a set of d directions $e_1, ..., e_d$. When the e_i form the canonical basis of R^d indicating the d axes, they reproduce the standard algorithms presented so far. When the e_i form another basis of R^d, the algorithm corresponds to sampling through components after a linear reparametrization. Other choices of direction are available. In particular, random choices can be made. Completely random choices correspond to the algorithm of Schmeiser and Chen (1991) observed at every d iterations. Phillips and Smith (1993) suggest only choosing e_1 at random and then choosing $e_2, ..., e_d$ to be mutually orthogonal and orthogonal to e_1. The directions can also be tuned to improve sampling in every specific setting. This will generally destroy the convergence properties of the chain so this tuning can only be operated at a transient phase of the chain.

Phillips and Smith (1993) also suggest a generalization in the choice of c. When sampling c from a general density $g(c|\theta^{(j-1)}, e^{(j)})$, the proposed value of the chain $\theta^{(j)} = \theta^{(j-1)} + c^{(j)} e^{(j)}$ is only accepted with probability

$$\min\left\{1, \frac{\pi(\theta^{(j-1)} + c^{(j)} e^{(j)}) g(-c^{(j)}|\theta^{(j-1)} + c^{(j)} e^{(j)}, e^{(j)})}{\pi(\theta^{(j-1)}) g(c^{(j)}|\theta^{(j-1)}, e^{(j)})}\right\}$$

The Schmeiser and Chen (1991) algorithm corresponds to $g(c|\theta^{(j-1)}, e^{(j)}) \propto \pi(\theta^{(j-1)} + c e^{(j)})$ and accepting the proposed value with probability 1.

These algorithms work well for the possibility of large moves when appropriate directions are chosen. This is generally not possible when moving along the direction of the components. The performance of these algorithms in highly dimensional models with correlated parameters or with concentrated modes may be very poor. The empirical evidence obtained by Phillips and Smith (1993) suggest that the orthogonalization of random directions improves convergence over sampling from entirely random directions which also improves convergence over Gibbs sampling. The improvement increases substantially when directions are chosen according to a principal component analysis based on (an approximation to) the posterior variance matrix. The comparisons mentioned above were based on the convergence prescription of Raftery and Lewis (1992) and on the integrated autocorrelation time (Green and Han, 1992).

Another approach to the appropriate selection of the sampling directions is given in Gilks, Roberts and George (1994). They proposed adaptive methods for choosing the direction of sampling. The method is based on a current sample $\theta_1^{(j)}, ..., \theta_n^{(j)}$ at each iteration of the chain and allows the choice of the next direction $e^{(j+1)}$ to be based on the current sample. Dependence on a succesion of values is not a problem for convergence as the Markov chain may be enlarged to contemplate n of the original steps. At each iteration, an element from the sample is randomly chosen and replaced by a point chosen according to a specified direction. This direction may depend on the other points in the current sample. When $n = 1$, the Schmeiser and Chen (1991) algorithm is obtained. A variation is to allow only directions connecting the chosen point to the other points in the sample. When these choices are made with the posterior controlling the displacements along the directions, the points are automatically chosen. When the displacements are chosen according to some other distribution, the points are only proposed and an acceptance probability must be evaluated.

The motivation for the adaptive methods is that the sample points will cover adequately regions of high probability of π near the equilibrium and subsequent points will tend to concentrate in these areas. For slow mixing chains, however, the sample from initial iterations may be far from these regions. Movements of the sample towards high probability regions may be slow. Hence, these difficulties and the extra computations required prevent a straightforward recommendation of these techniques. Rather, they should be used as auxiliary devices.

Finally, convergence acceleration can also be achieved by directly changing the chain output with resampling techniques. Assume m parallel chains are used for sampling and a sample $\theta_1^{(j)}, ..., \theta_m^{(j)}$ is available at iteration j. This is a sample from the marginal distribution of the chain $\pi^{(j)}$ at iteration j and the objective is to use this sample to obtain a sample from π. Successively drawing samples through the chain will attain that but Gelfand and Sahu (1994) consider the possibility of making this sample an approximate sample from π in a single iteration. Methods to achieve this goal were described in section 1.5 where it was shown how an arbitrary approximating density q can be used to generate a sample from π. In the present context, the approximating density $\pi^{(j)}$ is not directly available and must be estimated from its sample. Methods for doing it were discussed in Chapter 5. Once this density estimate $\hat{\pi}^{(j)}$ is obtained, resampling techniques can be used. Rejection methods are preferred to weighted resampling in principle when calculation of the enveloping constant is feasible. This is rarely the case and Gelfand and Sahu (1995) concentrate on weighted resampling.

An approximate sample from π is obtained by forming weights $w_i \propto \pi(\theta_i^{(j)})/\hat{\pi}^{(j)}(\theta_i^{(j)})$, $i = 1, ..., m$, and resampling from the discrete distribution with probabilities $w_1, ..., w_m$. If $\hat{\pi}^{(j)}$ is a reasonable approximation for π and m is large, the resulting resample will be a good approximation to a sample from π. This resample substitutes the current sample with consequent approximation to the limiting distribution. Even when the approximation is not very accurate, the resulting resample should be closer to a sample from π with convergence improvements.

Another suggestion made by Gelfand and Sahu (1995) is the adaptation of the transition kernels to forms that speed up the convergence. They work in the discrete case with an adaptive initial phase in the chain designed to identify better transition matrices among those with the same limiting distribution. This procedure does not affect the convergence of the chain but allows identification of better transitions. The changes between transitions in the initial phase are deterministic and do not violate convergence results. They apply these results to the choice of tuning parameters in specific proposal transition kernels, thus providing further theoretical justification for their choice (see also section 6.3).

7.4.2 Alterations to the equilibrium distribution

The use of resampling described above was applied to intermediate iterations of the chain, before it had reached equilibrium. The same ideas could be applied directly to the equilibrium stage. Assume it is possible to construct a chain with fast convergence to a limiting distribution q that approximates π. A sample from q is available after convergence of the chain is established but the objective is to obtain a sample from π. Once again, resampling methods can be used. Unlike the previous case, the expression for q will typically be available here but the enveloping constant will still be difficult to obtain. Weighted resampling seems a better option although it provides a sample from an approximation to π. Weights w_i based on the ratio π/q at sampled values can be calculated and the resample drawn from the discrete distribution concentrated at the sample from q with respective weights $w_1, ..., w_m$.

One possibility for q is obtained by *heating* the target distribution according to $q(\theta) \propto \pi(\theta)^{1/T}$ where the constant $T > 1$ receives the physical interpretation of system temperature, hence the nomenclature used. This mechanism was suggested by Jennison (1993) and is based on the optimization technique known as simulated annealing (Kirkpatrick, Gelatt and Vecchi, 1983; Ripley, 1987). The equilibrium distribution π corresponds to the basal temperature $T = 1$. The *heated* distribution q is flattened with respect to π and its density gets closer to the uniform distribution. It becomes easier to design a Markov chain that converges faster to the equilibrium distribution. This alteration is particularly relevant for the case of a distribution with distant modes. It is generally difficult to construct chains allowing for frequent moves between regions around the modes. Gibbs sampling will make these movements very slowly and Metropolis-Hastings algorithms will tend to reject most of these moves. By flattening the modes, the moves required to cover adequately the parameter space become more likely. Gilks and Roberts (1996) suggest modifying the distribution by the inclusion of what they call *stepping stones*. These are lumps of probability redistributed to regions to ease the moves between modes. This is in a way a discrete version of the *heating* procedure that redistributes weights continuously over the parameter space. Besag and Green (1993) also discussed approaches to multimodality in more qualitative terms.

The idea of *heated* equilibrium distributions was also used by Geyer (1991). The difference here is to use m parallel chains, each having equilibrium distribution

$$\pi_i(\theta) \propto \pi(\theta)^{1/T_i} \qquad (7.11)$$

gradually *heated* according to the rule $T_i = 1 + \lambda(i - 1)$, $i = 1, ..., m$, for some $\lambda > 0$. The target distribution is simply one of these chains, corresponding to the case $i = 1$. Chains with higher temperatures will have more movement than *cooler* chains, including the chain of interest $i = 1$. The objective here is to make the low temperature chain benefit from the moves from the *heated* chains. So, jumps between the chains are proposed in addition to regular moves within each chain. At iteration j, an exchange between the states of chains i and k is proposed. The acceptance probability of this swap is

$$\alpha_{ik}(\theta_i^{(j)}, \theta_k^{(j)}) = \min\left\{1, \frac{\pi_i(\theta_k^{(j)})\pi_k(\theta_i^{(j)})}{\pi_i(\theta_i^{(j)})\pi_k(\theta_k^{(j)})}\right\}$$

After convergence is reached at the m chains, a sample from chain $i = 1$ is drawn. The remaining $m - 1$ are only used to speed the convergence of the chain of interest and once that is achieved, they are discarded. This is an obvious computational disadvantage of the method. Another disadvantage is that sampling from the heated distributions will generally be more difficult.

A similar idea was proposed by Marinari and Parisi (1992) under the name of simulated tempering. Again, m chains with respective equilibrium distributions π_i, $i = 1, ..., m$, are used, but in series and not in parallel as above. One possibility for the π_i is given by (7.11). The chain thus formed alternates between equilibrium distributions and is formally extended to a chain of (θ, i) where i denotes the sampling scheme (with respective equilibrium distribution π_i) considered.

This structure is very similar to that used for model choice based on chains with jumps. The only changes are the fact that the parameter is the same through all components and the interpretation given to the additional component introduced. For model choice, it represents the model considered. Here, it only specifies the auxiliary chain constructed to accelerate the convergence of the chain of interest. Geyer and Thompson (1995) suggest that only jumps between neighbouring models are allowed and provide further discussion on implementation issues. Once again, after convergence is reached, only samples of θ corresponding to $i = 1$ are retained.

7.4.3 Auxiliary variables

Simulated tempering is a scheme where an additional variable was introduced with the aim of reducing slow mixing of the original chain. Slow mixing of the chain leads to slow convergence and is normally due to particularities of the target distribution such as multimodality or high correlation between some of the components of θ. Whatever the reason, convergence is slow due to high autocorrelation in the chain.

The introduction of auxiliary variables attempts to remove these sources of correlation and to hasten convergence of the extended chain. Consider the introduction of variables ϕ with known (and preferably easy to sample from) conditional distribution $\pi(\phi|\theta)$. The equilibrium distribution now becomes $\pi(\theta, \phi) = \pi(\theta)\pi(\phi|\theta)$ where the first term to the right hand side is the target distribution. The extended chain alternates generations from the full conditional distributions $\pi_\theta(\theta) \propto \pi(\theta)\pi(\phi|\theta)$ and $\pi_\phi(\phi) = \pi(\phi|\theta)$ at every iteration. If these generations can be made directly, Gibbs sampling can be applied to blocks θ and ϕ. Besag and Green (1993) consider chains with more general transition kernels.

Example 7.6 Assume that θ has a posterior distribution that can be written as

$$\pi(\theta) = q(\theta) \prod_{i=1}^{I} b_i(\theta)$$

where q has an easy to sample distribution and the functions b_i are complicated terms involving interactions between the components of the vector θ. Examples include space and temporal correlation. A vector $\phi = (\phi_1, ..., \phi_I)'$ of components conditionally independent given θ can be defined with distributions $\phi_i|\theta \sim U[0, b_i(\theta)]$, $i = 1, ..., I$. Generation from $\phi|\theta$ is simple and

$$\begin{aligned}
\pi(\theta, \phi) &= \pi(\theta)\,\pi(\phi|\theta) \\
&= q(\theta) \prod_{i=1}^{I} b_i(\theta) \times \prod_{i=1}^{I} \frac{I(0 \le \phi_i \le b_i(\theta))}{b_i(\theta)} \\
&= q(\theta) \prod_{i=1}^{I} I(0 \le \phi_i \le b_i(\theta))
\end{aligned}$$

Generation of θ from π_θ involves a generation from q followed by verification of conditions $b_i(\theta) \ge \phi_i$. By the rejection method, the generated value is retained if they are all satisfied. Otherwise, a new value is generated from q until all conditions are satisfied.

Example 7.7 The vector $\theta = (\theta_1, ..., \theta_d)'$ represents the colours in the pixels of a given image. Each position i has colour θ_i varying in a finite set of possibilities $\{1, ..., L\}$. A frequently adopted distribution for θ is given by the Potts model that seeks to reflect similarities in colours of neighbouring pixels. Its probability function is given by the Gibbs distribution with energy $E(\theta)$ being the number of neighbouring pairs of the same colour. So, $E(\theta) = \sum_{j \sim k} I(\theta_j \ne \theta_k)$, where $j \sim k$ denotes that the pair $i = (j, k)$ consists of neighbouring positions. The probability function can be written as

$$\pi(\theta) \propto \prod_i b_i(\theta) \quad \text{where } b_i(\theta) = e^{-\beta I(\theta_j \ne \theta_k)} \text{ and } \beta = 1/(kT)$$

If β is large, there is high correlation between the components of θ and the use of auxiliary variables is recommended.

Define an auxiliary vector $\phi = (\phi_1, ..., \phi_I)$ where I is the number of pairs of neighbours and the auxiliary variable ϕ_i associated with the pair $i = (j, k)$ have independent $bern(b_i(\theta))$ distributions. If $\theta_j = \theta_k$, $\phi_i = 1$ and if $\theta_j \neq \theta_k$, $\phi_i \sim bern(e^{-\beta})$. Generation from $\phi|\theta$ is therefore trivial. Generation of $\theta|\phi$ is based on a uniform distribution over $\{1, ..., L\}^d$. The generated value will be accepted if all pairs of neighbours satisfy the configuration given by ϕ. This generation mechanism was proposed by Swendsen and Wang (1987). It is discussed in the statistical context by Besag and Green (1993) and Green (1996).

Gilks and Roberts (1996) consider other possible uses of auxiliary variables. They include the important case of missing observations and a version of the rejection method where the auxiliary variable is again an indicator variable controlling the acceptance probability. Gilks, Best and Tan (1995) suggested a generalization of the adaptive rejection method where the acceptance probability is replaced by a Metropolis step. They show that their method can also be seen as another use of auxiliary variables. Finally, Besag and Green (1993) discuss other possibilities still preserving reversibility of the chain.

7.5 Exercises

1. Show that all the estimators of the model likelihood presented in section 7.2 are consistent and obtain their asymptotic variance. Also, discuss whether variance of $\hat{f}_4(y)^{-1}$ diverges or not.

2. Show that the densities used in the intrinsic Bayes factor of Berger and Pericchi (1996) and the fractional Bayes factor of O'Hagan (1995) can be written in the form $p(y_{S_1}|y_{S_2})$, identifying the sets S_1 and S_2 in each case.

3. Consider the model choice setting of section 7.3 with a superparameter $\theta = (\theta_1, ..., \theta_J)$ and a quantity M indicating a model with joint posterior $\pi(\theta, j)$.

   (a) Obtain the posterior distributions of $\theta_j|M = j$, $j = 1, ..., J$, and M. Show that samples from $\theta_j|M = j$ are obtained by retaining the draws of θ_j associated with a value j for M.

   (b) Consider a component of the parameter, say ϕ, that is shared by all models. Obtain its marginal posterior distribution $\pi(\phi)$.

4. Show that the supermodel approach of Carlin and Chib (1995) cannot be applied when the number of models considered is not finite.

5. Show that the reversible jump Markov chain approach for inference with a collection of models reduces to the Metropolis-Hastings algorithm if

only one model is considered and moves are always proposed. What happens if there is a positive probability of not proposing a move?

6. Consider the reversible jump Markov chain approach with transition kernel $q_m((\theta, j), (\phi, k))$ for each proposed move m. Show that reversibility of chain is ensured if the acceptance probability associated with move m is given by

$$\alpha_m((\theta, j), (\phi, k)) = \min\left\{1, \frac{\pi(\phi, k)q_m((\phi, k), (\theta, j))}{\pi(\theta, j)q_m((\theta, j), (\phi, k))}\right\}$$

7. (Green, 1995) Consider again the conditions of Example 7.4 with irreducible birth and death chains moving across models M_k with piecewise constant intensity rates having k steps and model parameter $\theta_k = (\lambda(k), t(k))'$ where $\lambda(k) = (\lambda_0, \lambda_1, ..., \lambda_k)$ and $t(k) = (t_1, ..., t_k)$, $k = 0, 1, 2, ...$

   (a) Show that the likelihood for model M_k is given by

   $$l(\theta_k, k) = \prod_{j=0}^{k} \lambda_j^{d_j} e^{-\lambda_j(t_{j+1}-t_j)}$$

   where d_j is the number of occurrences in interval I_j, $j = 0, 1, ..., k$.

   (b) Show that the likelihood ratio between model M_k with parameters θ'_k and θ_k is given by

   $$\mathrm{lr}_\lambda = \prod_{j=0}^{k} \left(\frac{\lambda'_j}{\lambda_j}\right)^{d_j} e^{(\lambda_j - \lambda'_j)(t_{j+1}-t_j)}$$

   if they differ only on the values of $\lambda(k)$ and

   $$\mathrm{lr}_t = \prod_{j=0}^{k} \lambda_j^{d'_j - d_j} e^{\lambda_j[(t_{j+1}-t_j)-(t'_{j+1}-t'_j)]}$$

   if they differ only on the values of $t(k)$.

   (c) Consider a move of type 3 which proposes a change of the intensity rate of a randomly chosen interval, say j, from λ_j to λ'_j according to a random walk $log \, \lambda'_j \sim U[\lambda_j - 1/2, \lambda_j + 1/2]$. Show that the acceptance probability of this move is

   $$\min\left\{1, lr_\lambda \left(\frac{\lambda'_j}{\lambda_j}\right)^\alpha \exp[-\beta(\lambda'_j - \lambda_j)]\right\}$$

   (d) Consider a move of type 4 which proposes a change of the endpoint of a randomly chosen interval, say j, from t_j to t'_j according to a $U[t_{j-1}, t_{j+1}]$ distribution. Show that the acceptance probability of this move is

$$\min\left\{1, \text{lr}_t \frac{(t_{j+1} - t'_j)(t'_j - t_{j-1})}{(t_{j+1} - t_j)(t_j - t_{j-1})}\right\}$$

(e) Consider a move of type 1 which proposes the birth of a new end-point at a point uniformly chosen on $[0, T]$. Show that the acceptance probability of this move is

$$\min\left\{1, \frac{\pi(\theta_{k+1}, k+1)}{\pi(\theta_k, k)} \times \frac{B_{k+1,k}}{B_{k,k+1} q(t^*, u)} \times \left|\frac{\partial\theta_{k+1}}{\partial(\theta_k, t^*, u)}\right|\right\}$$

(f) Show that in the above expression

$$\frac{\pi(\theta_{k+1}, k+1)}{\pi(\theta_k, k)} = \frac{l(\theta_{k+1}, k+1)}{l(\theta_k, k)} \times \frac{p(\lambda(k+1), t(k+1), k+1)}{p(\lambda(k), t(k), k)}$$

where

$$p(\lambda(l), t(l), l) = p(\lambda(l)|l)\, p(t(l)|l)\, f_P(l)\,, \; l = 1, 2, \ldots,$$

$$\frac{p(\lambda(k+1)|k+1)}{p(\lambda(k)|k)} = \frac{\beta^\alpha}{\Gamma(\alpha)}\left(\frac{\lambda'_j\lambda'_{j+1}}{\lambda_j}\right)^{\alpha-1} e^{-\beta(\lambda'_j + \lambda'_{j+1} - \lambda_j)} \text{ and}$$

$$\frac{p(t(k+1)|k+1)}{p(t(k)|k)} = \frac{2(k+1)(2k+3)}{T^2}\frac{(t^* - t_j)(t_{j+1} - t^*)}{t_{j+1} - t_j}$$

(g) Show that $B_{k,k+1} = p_k$, $B_{k+1,k} = q_{k+1}/(k+1)$ and $q(t^*, u) = 1/T$ and that the Jacobian is given by $(\lambda'_j + \lambda'_{j+1})^2/\lambda_j$.

(h) Show that the acceptance probability of a death move is

$$\min\left\{1, \frac{\pi(\theta_k, k)}{\pi(\theta_{k+1}, k+1)} \times \frac{B_{k+1,k} q(t^*, u)}{B_{k+1,k}} \times \left|\frac{\partial(\theta_k, t^*, u)}{\partial\theta_{k+1}}\right|\right\}$$

8. Show that the Schmeiser and Chen (1991) algorithm corresponds to taking the proposal density g in the Phillips and Smith (1993) algorithm as $g(c|\theta^{(j-1)}, e^{(j)}) \propto \pi(\theta^{(j-1)} + ce^{(j)})$. Discuss other possible forms for g, commenting on their advantages/disadvantages with respect to the above choices.

9. Generate samples from the Potts model described in Example 7.6 varying the values of β and using both a componentwise Markov chain sampler and the method of auxiliary variables. Compare the convergence of the generated samples from both approaches.

References

Abramowitz, M. and Stegun, I. A. (eds) (1965) *Handbook of Mathematical Functions*, National Bureau of Standards, Washington.

Achcar, J. A. and Smith, A. F. M. (1989) Aspects of reparametrisation in approximate Bayesian inference, in *Bayesian and Likelihood Methods in Statistics and Econometrics: Essays in Honour of George A. Barnard* (eds S. Geisser et al.), North-Holland, Amsterdam, pp. 439-52.

Ahrens, J. H. and Dieter, U. (1974) Computer methods for sampling gamma, beta, Poisson and binomial distributions. *Computing*, **12**, 223-46.

Aitchinson, J. and Dunsmore, I. R. (1975) *Statistical Prediction Analysis*, Cambridge University Press, Cambridge.

Aitkin, M. (1991) Posterior Bayes factors (with discussion). *Journal of the Royal Statistical Society, Series B*, **53**, 111-42.

Albert, J. H. (1988) Computational methods using a Bayesian hierarchical generalized linear model. *Journal of the American Statistical Association*, **83**, 1037-45.

Albert, J. H. (1996) A MCMC algorithm to fit a general exchangeable model. *Communications in Statistics - Simulation and Computation*, **25**, 573-92.

Anderson, T. W. (1958) *An Introduction to Multivariate Statistical Analysis*, Wiley, New York.

Andrews, D. F. and Mallows, C. L. (1974) Scale mixtures of normality. *Journal of the Royal Statistical Society, Series B*, **36**, 99-102.

Arjas, E. and Heikkinen, J. (1997) An algorithm for nonparametric Bayesian estimation of a Poisson intensity. To appear in *Computational Statistics*.

Barker, A. A. (1965) Monte Carlo calculation of the radial distribution functions for a protonelectron plasma. *Australian Journal of Physics*, **18**, 119-33.

Barone, P. and Frigessi, A. (1989) Improving stochastic relaxation for Gaussian random fields. *Probability in the Engineering and Informational Sciences*, **4**, 369-89.

Bennett, J. E., Racine-Poon, A. and Wakefield, J. C. (1996) MCMC for nonlinear hierarchical models, in *Markov Chain Monte Carlo in Practice* (eds W. R. Gilks, S. Richardson and D. J. Spiegelhalter), Chapman & Hall, London, pp. 339-57.

Berger, J. O. (1985) *Statistical Decision Theory and Bayesian Analysis*, 2nd edn, Springer Verlag, New York.

Berger, J. O. and Bernardo, J. M. (1992) On the development of reference priors (with discussion), in *Bayesian Statistics 4* (eds J. M. Bernardo et al.), Oxford University Press, Oxford. pp. 35-60.

Berger, J. O., Bernardo, J. M. and Mendoza, M. (1989) On priors that maximize expected information, in *Recent Development in Statistics and their Applications* (eds J. P. Klein and J. C. Lee), Freedom Academy Publishing: Seoul, pp. 1-20.

Berger, J. O. and Pericchi, L. R. (1996) The intrinsic Bayes factor for model selection and prediction. *Journal of the American Statistical Association*, **91**, 109-22.

Bernardo, J. M. (1979) Reference posterior distributions for Bayesian inference (with discussion). *Journal of the Royal Statistical Society, Series B*, **41**, 113-47.

Besag, J. (1974) Spatial interaction and the statistical analysis of life systems (with discussion). *Journal of the Royal Statistical Society, Series B*, **48**, 192-236.

Besag, J. (1986) On statistical analysis of dirty pictures (with discussion). *Journal of the Royal Statistical Society, Series B*, **48**, 259-302.

Besag, J. and Green, P. J. (1993) Spatial statistics and Bayesian computation (with discussion). *Journal of the Royal Statistical Society, Series B*, **55**, 25-37.

Besag, J., Green, P. J., Higdon, D. and Mengersen, K. (1995) Bayesian computation and stochastic systems (with discussion). *Statistical Science*, **10**, 3-66.

Besag, J, York, J. and Mollié, A. (1991) Bayesian image restoration, with two applications in spatial statistics (with discussion). *Annals of the Institute of Statistical Mathematics*, **43**, 1-59.

Best, N. G., Cowles, M. K. and Vines, S. K. (1995) *CODA: Convergence Diagnostics and Output Analysis Software for Gibbs Sampler Output: Version 0.3*. Technical report, Biostatistics Unit-MRC, Cambridge, UK.

Box, G. E. P. and Muller, M. E. (1958) A note on the generation of random normal deviates. *Annals of Mathematical Statistics*, **29**, 610-1.

Breslow, N. E. and Clayton, D. (1993) Approximate inference in generalized linear mixed models. *Journal of the American Statistical Association*, **88**, 9-25.

Brooks, S. P. and Roberts, G. O. (1995) Diagnosing convergence of Markov chain Monte Carlo algorithms. Technical Report 95-12, Statistical Laboratory, University of Cambridge.

Capocaccia, D, Cassandro, M. and Olivieri, E. (1977) A study of metastability in the Ising model. *Communications in Mathematical Physics*, **39**, 185-205.

Carlin, B. P. and Chib, S. (1995) Bayesian model choice via Markov chain

Monte Carlo methods. *Journal of the Royal Statistical Society, Series B*, **57**, 473-84.

Carlin, B. P., Gelfand, A. E. and Smith, A. F. M. (1992) Hierarchical Bayesian analysis of changepoint problems. *Applied Statistics*, **41**, 389-405.

Carlin, B. P. and Louis, T. (1996) *Bayes and Empirical Bayes methods for Data Analysis*, Chapman & Hall, London.

Carlin, B. P., Polson, N. G. and Stoffer, D. S. (1992) A Monte Carlo approach to nonnnormal and nonlinear state-space modeling. *Journal of the American Statistical Association*, **87**, 493-500.

Carter, C. K. and Kohn, R. (1994) On Gibbs sampling for state space models. *Biometrika*, **81**, 541-53.

Casella, G. and George, E. I. (1992) Explaining the Gibbs sampler. *The American Statistician*, **46**, 167-74.

Cassandro, M., Galves, A., Olivieri, E. and Vares, M. E. (1984) Metastable behaviour of stochastic dynamics: a pathwise approach. *Journal of Statistical Physics*, **35**, 603-34.

Chambers, J. M. (1977) *Computational Methods for Data Analysis*, Wiley, New York.

Chan, K. S. and Geyer, C. J. (1994) Discussion of the paper by Tierney (1994). *Annals of Statistics*, **22**, 1747-58.

Cheng, R. C. H. and Feast, G. M. (1979) Some simple gamma variate generators. *Applied Statistics*, **28**, 290-5.

Chib, S. (1995) Marginal likelihood from the Gibbs output. *Journal of the American Statistical Association*, **90**, 773-95.

Chib, S. and Greenberg, E. (1994) Bayes inference for regression models with ARMA(p,q) models. *Journal of Econometrics*, **64**, 183-206.

Chib, S. and Greenberg, E. (1995) Understanding the Metropolis-Hastings algorithms. *The American Statistician*, **49**, 327-35.

Clayton, D. G. (1996) Generalized linear mixed models, in *Markov Chain Monte Carlo in Practice* (eds W. R. Gilks, S. Richardson and D. J. Spiegelhalter), Chapman & Hall, London, pp. 275-301.

Cowles, M. K. and Carlin, B. P. (1996) Markov chain Monte Carlo convergence diagnostics: a comparative review. *Journal of the American Statistical Association*, **91**, 883-904.

Cox, D. R. and Reid, N. (1987) Parameter orthogonality and approximate conditional inference (with discussion). *Journal of the Royal Statistical Society, Series B*, **49**, 1-39.

Crowder, M. J. (1978) Beta-binomial ANOVA for proportions. *Applied Statistics*, **27**, 34-7.

Dagpunar, J. (1988) *Principles of Random Variate Generation*, Clarendon Press, Oxford.

Deely, J. J. and Lindley, D. V. (1981) Bayes empirical Bayes. *Journal of the American Statistical Association*, **76**, 833-41.

De Finetti, B. (1974) *The Theory of Probability*, Vol. I, Wiley, Chichester.

De Finetti, B. (1975) *The Theory of Probability*, Vol. II, Wiley, Chichester.

DeGroot, M. H. (1970) *Optimal Statistical Decisions*, McGraw-Hill, New York.

DeGroot, M. H. (1986) *Probability and Statistics* (2nd edn), Addison Wesley, Reading.

Dellaportas, P. and Wright, D. (1991) Positive embedded integration in Bayesian analysis. *Statistics and Computing*, 1, 1-12.

Dellaportas, P. and Smith, A. F. M. (1993) Bayesian inference for generalized linear and proportional hazards models via Gibbs sampling. *Applied Statistics*, 42, 443-60.

Dennison, D., Mallick, B. and Smith, A. F. M. (1996), Bayesian CART models. Technical Report, Statistics Section, Imperial College London.

Devroye, L. (1986) *Non-uniform Random Variate Generation*, Springer Verlag, New York.

Efron, B. (1986) Why isn't everyone a Bayesian? *American Statistician*, 40, 1-11.

Elston, R. C. and Grizzle, J. E. (1962) Estimation of time-response curves and their confidence bands. *Biometrics*, 18, 148-59.

Evans, M. and Swartz, T. (1988) Sampling from Gauss rules. *SIAM Journal on Scientific and Statistical Computing*, 9, 950-61.

Evans, M. and Swartz, T. (1995) Methods for approximating integrals in statistics with special emphasis on Bayesian integration problems. *Statistical Science*, 10, 254-72.

Fahrmeir, L. and Wagenpfeil, S. (1997) Penalized likelihood estimation and iterative Kalman smoothing for non-gaussian dynamic regression models. *Computational Statistics and Data Analysis*, 24, 295-320.

Feller, W. (1968) *An Introduction to Probability Theory and Its Applications*, Volume I, 3rd edn, Wiley, New York.

Fishman, G. and Moore, L. (1985) An exhaustive analysis of multiplicative congruential random number generators with modulus $2^{31} - 1$. *SIAM Journal of Scientific and Statistical Computing*, 7, 24-45.

Fruhwirth-Schnatter, S. (1994) Data augmentation and dynamic linear models. *Journal of Time Series Analysis*, 15, 183-202.

Gamerman, D. (1997) Efficient sampling from the posterior distribution in generalized linear mixed models. *Statistics and Computing*, 7, 57-68.

Gamerman, D. (1998) Markov chain Monte Carlo for dynamic generalized linear models. To appear in *Biometrika*.

Gamerman, D. and Migon, H. S. (1993). *Statistical Inference: an Integrated Approach*, Monograph of the Institute of Mathematics, UFRJ (in Portuguese).

Gamerman, D. and Smith, A. F. M. (1996) Bayesian analysis of longitudinal data studies, in *Bayesian Statistics 5* (eds J. M. Bernardo et al.), Oxford University Press, Oxford, pp. 587-98.

Garren, S. and Smith, R. L. (1993) Convergence diagnostics for Markov chain samplers. Technical report, University of North Carolina.

Geisser, S. (1993) *Predictive Inference: an Introduction*, Chapman & Hall, London.

Geisser, S. and Eddy, W. (1979) A predictive approach to model selection. *Journal of the American Statistical Association*, **74**, 153-60.

Gelfand, A. E. (1992) Discussion of the paper by Gelman and Rubin. *Statistical Science*, **7**, 486-7.

Gelfand, A. E. (1996) Model determination using sampling-based methods, in *Markov Chain Monte Carlo in Practice* (eds W. R. Gilks, S. Richardson and D. J. Spiegelhalter), Chapman & Hall, London, pp. 145-61.

Gelfand, A. E. and Carlin, B. P. (1995) Discussion of the paper by Besag, Green, Higdon and Mengersen. *Statistical Science*, **10**, 43-6.

Gelfand, A. E. and Dey, D. K. (1994) Bayesian model choice: asymptotics and exact calculations. *Journal of the Royal Statistical Society, Series B*, **56**, 501-14.

Gelfand, A. E., Dey, D. K. and Chang, H. (1992) Model determination using predictive distributions with implementation via sampling-based methods, in *Bayesian Statistics 4* (eds J. M. Bernardo et al.), Oxford University Press, Oxford, pp. 147-67.

Gelfand, A. E., Hills, S. E., Racine-Poon, A. and Smith, A. F. M. (1990) Illustration of Bayesian inference in normal data models using Gibbs sampling. *Journal of the American Statistical Association*, **85**, 972-85.

Gelfand, A. E. and Sahu, S. K. (1994) On Markov chain Monte Carlo acceleration. *Journal of Computational and Graphical Statistics*, **3**, 261-7.

Gelfand, A. E., Sahu, S. K. and Carlin, B. P. (1995) Efficient parametrisations for normal linear mixed models. *Biometrika*, **82**, 479-488.

Gelfand, A. E., Sahu, S. K. and Carlin, B. P. (1996) Efficient parametrisations for generalized linear mixed models (with discussion), in *Bayesian Statistics 5* (eds J. M. Bernardo et al.), Oxford University Press, Oxford, pp. 165-80.

Gelfand, A. E. and Smith, A. F. M. (1990) Sampling-based approaches to calculating marginal densities. *Journal of the American Statistical Association*, **85**, 398-409.

Gelman, A. (1996) Inference and monitoring convergence, in *Markov Chain Monte Carlo in Practice* (eds W. R. Gilks, S. Richardson and D. J. Spiegelhalter), Chapman & Hall, London, pp. 131-43.

Gelman, A., Carlin, J. B., Stern, H. S. and Rubin, D. B. (1995) *Bayesian Data Analysis*, Chapman & Hall, London.

Gelman, A., Roberts, G. O. and Gilks, W. R. (1996). Efficient Metropolis jumping rules, in *Bayesian Statistics 5* (eds J. M. Bernardo et al.), Oxford University Press, Oxford, pp. 599-607.

Gelman, A. and Rubin, D. R. (1992a) A single series from the Gibbs sampler provides a false sense of security, in *Bayesian Statistics 4* (eds J. M. Bernardo et al.), Oxford University Press, Oxford, 625-31.

Gelman, A. and Rubin, D. R. (1992b) Inference from iterative simulation using multiple sequences (with discussion). *Statistical Science*, **7**, 457-511.

Geman, S. and Geman, D. (1984) Stochastic relaxation, Gibbs distributions and the Bayesian restoration of images. *IEEE Transactions on Pattern Analysis and Machine Intelligence*, **6**, 721-41.

George, E. I., Makov, U. E. and Smith, A. F. M. (1993) Conjugate likelihood distributions. *Scandinavian Journal of Statistics*, **20**, 147-156.

Geweke, J. (1989) Bayesian inference in econometric models using Monte Carlo integration. *Econometrica*, **57**, 1317-39.

Geweke, J. (1992) Evaluating the accuracy of sampling-based approaches to the calculation of posterior moments (with discussion), in *Bayesian Statistics 4* (eds J. M. Bernardo et al.), Oxford University Press, Oxford, pp. 169-93.

Geyer, C. J. (1991) Markov chain Monte Carlo maximum likelihood, in *Computing Science and Statistics: Proceedings of the 23rd. Symposium on the Interface* (ed. E. M. Keramidas), Fairfax Station: Interface Foundation, pp. 156-163.

Geyer, C. J. (1992) Practical Markov chain Monte Carlo (with discussion). *Statistical Science*, **7**, 473-511.

Geyer, C. J. and Thompson, E. A. (1995) Annealing Markov chain Monte Carlo with applications to ancestral inference. *Journal of the American Statistical Association*, **90**, 909-20.

Gilks, W. R. (1992) Derivative-free adaptive rejection sampling for Gibbs sampling, in *Bayesian Statistics 4* (eds J. M. Bernardo et al.), Oxford University Press, Oxford, pp. 641-9.

Gilks, W. R., Best, N. G. and Tan, K. K. C. (1995) Adaptive rejection Metropolis sampling within Gibbs sampling. *Applied Statistics*, **44**, 455-72.

Gilks, W. R., Richardson, S. and Spiegelhalter, D. J. (eds) (1996) *Markov Chain Monte Carlo in Practice*, Chapman & Hall, London.

Gilks, W. R. and Roberts, G. O. (1996) Strategies for improving MCMC, in *Markov Chain Monte Carlo in Practice* (eds W. R. Gilks, S. Richardson and D. J. Spiegelhalter), Chapman & Hall, London, pp. 89-114.

Gilks, W. R., Roberts, G. O. and George, E. I. (1994) Adaptive direction sampling. *The Statistician*, **43**, 179-89.

Gilks, W. R. and Wild, P. (1992) Adaptive rejection sampling for Gibbs sampling. *Applied Statistics*, **41**, 337-48.

Gillespie, D. T. (1992) *Markov Processes: An Introduction for Physical Scientists*, Academic Press.

Gordon, N. J., Salmond, D. J. and Smith, A. F. M. (1993) Novel approach to nonlinear/non-Gaussian Bayesian state estimation. *IEE Proceedings F*, **140**, 107-13.

Green, P. J. (1991) Discussion of the paper by Besag, York and Mollié. *Annals of the Institute of Statistical Mathematics*, **43**, 22-4.

Green, P. J. (1995) Reversible jump Markov chain Monte Carlo computation and Bayesian model determination. *Biometrika*, **82**, 711-32.

Green, P. J. (1996) MCMC in image analysis, in *Markov Chain Monte Carlo in Practice* (eds W. R. Gilks, S. Richardson and D. J. Spiegelhalter), Chapman & Hall, London, pp. 381-99.

Green, P. J. and Han, X.-L. (1992) Metropolis methods, Gaussian proposals and antithetic variables, in *Lecture Notes in Statistics* **74**: Stochastic Models (eds A. Barone, A. Frigessi and M. Piccioni), Springer Verlag, pp. 142-64.

Grenander, U. and Miller, M. I. (1994) Representation of knowledge in complex systems (with discussion). *Journal of the Royal Statistical Society, Series B*, **56**, 549-603.

Guttorp, P. (1995) *Stochastic Modeling of Scientific Data*, Chapman & Hall, London.

Hammersley, J. M. and Handscomb, D. C. (1964) *Monte Carlo Methods*, Methuen, London.

Harrison, P. J. and Stevens, C. F. (1976) Bayesian forecasting (with discussion). *Journal of the Royal Statistical Society, Series B*, **38**, 205-47.

Hastings, W. K. (1970) Monte Carlo sampling methods using Markov chains and their applications. *Biometrika*, **57**, 97-109.

Heidelberger, P. and Welch, P. D. (1983) Simulation run length control in the presence of an initial transient. *Operations Research*, **31**, 1109-44.

Heyde, C. C. and Johnstone, I. M. (1979) On asymptotic posterior normality of stochastic processes. *Journal of the Royal Statistical Society, Series B*, **41**, 184-9.

Hills, S. E. and Smith, A. F. M. (1992) Parametrization issues in Bayesian inference (with discussion), in *Bayesian Statistics 4* (eds J. M. Bernardo et al.), Oxford University Press, Oxford, pp. 227-46.

Hoel, P. G., Port, S. C. and Stone, C. J. (1972) *Introduction to Stochastic Processes*, Houghton Mifflin.

Jacquier, E., Polson, N. G. and Rossi, P. E. (1994) Bayesian analysis of stochastic volatility models (with discussion). *Journal of Business and Economic Statistics*, **12**, 371-415.

Jeffreys, A. (1961) *The Theory of Probability*, Cambridge University Press, Cambridge.

Jennison, C. (1993) Discussion of the meeting on Gibbs sampling and other Markov chain Monte Carlo methods. *Journal of the Royal Statistical Society, Series B*, **55**, 54-6.

Johnson, V. E. (1996) Studying convergence of Markov chain Monte Carlo algorithms using coupled sample paths. *Journal of the American Statistical Association*, **91**, 154-66.

Kass, R. E. and Raftery, A. E. (1995) Bayes factors. *Journal of the American Statistical Association*, **90**, 773-95.

Kass, R. E. and Slate, E. H. (1992) Reparametrization and diagnostics of posterior nonnormality (with discussion), in *Bayesian Statistics 4* (eds J. M. Bernardo et al.), Oxford University Press, Oxford, pp. 289-305.

Kass, R. E. and Steffey, D. (1989) Approximate Bayesian inference in conditionally independent hierarchical models (parametric empirical Bayes models). *Journal of the American Statistical Association*, **84**, 717-26.

Kass, R. E., Tierney, L. and Kadane, J. B. (1988) Asymptotic in Bayesian computation (with discussion), in *Bayesian Statistics 3* (eds J. M. Bernardo et al.), Oxford University Press, Oxford, pp. 261-78.

Kinderman, A. J. and Monahan, J. F. (1977) Computer generation of random variables using the ratio of random deviates. *ACM Transactions in Mathematical Software*, **3**, 257-60.

Kirkpatrick, S, Gelatt, C. D., Jr. and Vecchi, M. P. (1983) Optimization by simulated annealing. *Science*, **220**, 671-80.

Kloek, J. and van Dijk, H. K. (1978) Bayesian estimates of equation system parameters: an application of integration by Monte Carlo. *Econometrica*, **46**, 1-19.

Knorr-Held, L. (1997) *Hierarchical Modelling of Discrete Longitudinal Data: Applications of Markov chain Monte Carlo*, Herbert Utz Verlag, Munich.

Laird, N. M. and Ware, J. H. (1982) Random effects models for longitudinal data. *Biometrics*, **38**, 963-74.

Laplace, P. S. (1986) Memoir on the probability of causes of events (translation by S. Stigler). *Statistical Science*, **1**, 364-78.

Lehmer, D. H. (1951) Mathematical methods in large-scale computing units, in *Proceedings of the Second Symposium on Large-Scale Digital Calculating Machinery*, Harvard University Press, Cambridge, pp. 141-6.

Lewis, P. A. W., Goodman, A. S. and Miller, J. M. (1969) A pseudo-random number generatot for the System 360. *IBM Systems Journal*, **8**, 136-45.

Lindley, D. V. (1961) The use of prior probability distributions in statistical inference and decison, in *Proceedings of the Fourth Berkeley Symposium on Mathematical Statistics and Probability*, **1**, 453-68.

Lindley, D. V. (1978) The Bayesian approach. *Scandinavian Journal of Statistics*, **5**, 1-26.

Lindley, D. V. (1980) Approximate Bayesian methods (with discussion), in *Bayesian Statistics* (eds J. M. Bernardo et al.), Valencia University Press, Valencia, pp. 223-45.

Lindley, D. V. and Smith, A. F. M. (1972) Bayes estimates for the linear model (with discussion). *Journal of the Royal Statistical Society, Series B*, **34**, 1-41.

Liu, C., Liu, J. and Rubin, D. R. (1992) A variational control variable for assessing the convergence of the Gibbs sampler, in *Proceedings of the American Statistical Association, Statistical Computing Section*, pp. 74-8.

Liu, J., Wong, W. H. and Kong, A. (1994) Correlation structure and convergence rate of the Gibbs sampler: applications to the comparison of estimators and augmentation schemes. *Biometrika*, **81**, 27-40.

MacEachern, S. N. and Berliner, L. M. (1994) Subsampling the Gibbs sampler. *The American Statistician*, **48**, 188-90.

Marinari, E. and Parisi, G. (1992) Simulated tempering: a new Monte Carlo scheme. *Europhysics Letters*, **19**, 451-8.

Marsaglia, G. (1977) The squeeze method for generating gamma variates. *Computers and Mathematics with Applications*, **3**, 321-5.

Marsaglia, G. and Bray, T. A. (1964). A conventional method for generating normal variables. *SIAM Review*, **6**, 260-4.

McCullagh, P. and Nelder, J. A. (1988) *Generalized Linear Models*, 2nd edn, Chapman & Hall, New York.

McCulloch, R. E. and Rossi, P. E. (1991) A Bayesian approach to testing the arbitrage pricing theory. *Journal of Econometrics*, **49**, 141-68.

Medhi, J. (1994) *Stochastic Processes*, 2nd edn, Wiley, New York.

Mendes, B. V. M. (1995) Bayesian inference using S-PLUS: the sampling-importance resampling technique. Technical report, Statistical Laboratory, UFRJ.

Meng, X. L. and Wong, W. H. (1993) Simulating ratios of normalizing constants via a simple identity. Technical Report, Statistics Department, University of Chicago, USA.

Metropolis, N., Rosenbluth, A. W., Rosenbluth, M. N., Teller, A. H. and Teller, E. (1953) Equation of state calculations by fast computing machine. *Journal of Chemical Physics*, **21**, 1087-91.

Meyn, S. P. and Tweedie, R. L. (1993) *Markov Chains and Stochastic Stability*, Springer, New York.

Meyn, S. P. and Tweedie, R. L. (1994) Computable bounds for convergence rates of Markov chains. *Annals of Applied Probability*, **4**, 124-48.

Migon, H. S. and Harrison, P. J. (1985) An application of non-linear Bayesian forecasting to television advertising, in *Bayesian Statistics 2*, (eds J. M. Bernardo et al.), North Holland, Amsterdam, pp. 681-96.

Mood, A. M., Graybill, F. A. and Boes, D. C. (1974) *Introduction to the Theory of Statistics*, 3rd edn, McGraw-Hill, Tokyo.

Moreira, A., Lopes, H. F. and Schmidt, A. M. (1996) Hyperparameter estimation in forecasting models. Technical report, IPEA, Brazil.

Muller, P. (1991a) Monte Carlo integration in general dynamic models. *Contemporary Mathematics*, **115**, 145-63.

Muller, P. (1991b) Metropolis based posterior integration schemes. Technical Report, Statistics Department, Purdue University.

Naylor, J. C. and Smith, A. F. M. (1982) Application of a method for the efficient computation of posterior distributions. *Applied Statistics*, **31**, 214-25.

Newman, T. G. and Odell, P. L. (1971) *The Generation of Random Variates*, Griffin, London.

Newton, M. A. and Raftery, A. E. (1994) Approximate Bayesian inference by the weighted likelihood bootstrap (with discussion). *Journal of the Royal Statistical Society, Series B*, **56**, 3-48.

Nummelin, E. (1984) *General Irreducible Markov Chains and Non-negative Operators*, Cambridge University Press, Cambridge.

Odell, P. L. and Feiveson, A. G. (1966) A numerical procedure to generate a sample covariance matrix. *Journal of the American Statistical Association*, **61**, 199-203.

Oh, M.-S. and Berger, J. O. (1992) Adaptive importance sampling in Monte Carlo integration. *Journal of Statistical Computation and Simulation*, **41**, 143-68.

Oh, M.-S. and Berger, J. O. (1993) Integration of multimodal functions by Monte Carlo importance sampling. *Journal of the American Statistical Association*, **88**, 450-6.

O'Hagan, A. (1987) Monte Carlo is fundamentally unsound. *The Statistician*, **36**, 247-9.

O'Hagan, A. (1994) *Bayesian Inference*, Volume 2B of Kendall's Advanced Theory of Statistics, Edward Arnold, London.

O'Hagan, A. (1995) Fractional Bayes factors (with discussion). *Journal of the Royal Statistical Society, Series B*, **57**, 99-138.

Peskun, P. H. (1973) Optimum Monte Carlo sampling using Markov chains. *Biometrika*, **60**, 607-12.

Phillips, D. B. and Smith, A. F. M. (1993) Orthogonal random-direction sampling in Markov chain Monte Carlo. Technical Report 93-06, Statistics Section, Imperial College London.

Phillips, D. B. and Smith, A. F. M. (1996) Bayesian model comparison via jump diffusions, in *Markov Chain Monte Carlo in Practice* (eds W. R. Gilks, S. Richardson and D. J. Spiegelhalter), Chapman & Hall, London, pp. 215-39.

Polson, N. G. (1996) Convergence of Markov chain Monte Carlo algorithms (with discussion), in *Bayesian Statistics 5* (eds J. M. Bernardo et al.). Oxford University Press, Oxford, pp. 297-321.

Press, S. J. (1989) *Bayesian Statistics: Principles, Models, and Applications*, Wiley, New York.

Priestley, M. B. (1981) *Spectral Analysis and Time Series*, Academic Press, London.

Raftery, A. E. (1996) Hypothesis testing and model selection, in *Markov Chain Monte Carlo in Practice* (eds W. R. Gilks, S. Richardson and D. J. Spiegelhalter), Chapman & Hall, London, pp. 165-87.

Raftery, A. E. and Lewis, S. (1992) How many iterations in the Gibbs sampler?, in *Bayesian Statistics 4* (eds J. M. Bernardo et al.), Oxford University Press, Oxford, pp. 763-73.

Raftery, A. E. and Lewis, S. (1996) Implementing MCMC, in *Markov Chain Monte Carlo in Practice* (eds W. R. Gilks, S. Richardson and D. J. Spiegelhalter), Chapman & Hall, London, pp. 115-30.

Ratkowski, D. A. (1983) *Nonlinear Regression Modeling: a Unified Practical Approach*, Marcel Dekker, New York.

Revuz, D. (1975) *Markov Chains*, North-Holland, Amsterdam.

Richardson, S. and Green, P. J. (1997) On Bayesian analysis of mixtures with an unknown number of components (with discussion). *Journal of the Royal Statistical Society, Series B*, **59**, 731-92.

Ripley, B. D. (1987) *Stochastic Simulation*, Wiley, New York.

Ritter, C. and Tanner, M. A. (1992) Facilitating the Gibbs sampler: the Gibbs stopper and the griddy Gibbs sampler. *Journal of the American Statistical Association*, **87**, 861-8.

Robert, C. P. (1994) *The Bayesian Choice: a Decision-Theoretic Motivation*, Springer Verlag, New York.

Robert, C. P. (1995) Convergence control methods for Markov chain Monte Carlo algorithms. *Statistical Science*, **10**, 231-53.

Roberts, G. O. (1992) Convergence diagnostics of the Gibbs sampler, in *Bayesian Statistics 4* (eds J. M. Bernardo et al.), Oxford University Press, Oxford, pp. 775-82.

Roberts, G. O. (1996) Markov chain concepts related to sampling algorithms, in *Markov Chain Monte Carlo in Practice* (eds W. R. Gilks, S. Richardson and D. J. Spiegelhalter), Chapman & Hall, London, pp. 45-57.

Roberts, G. O. and Polson, N. G. (1994) A note on the geometric convergence of the Gibbs sampler. *Journal of the Royal Statistical Society, Series B*, **56**, 377-84.

Roberts, G. O. and Sahu, S. K. (1997) Updating schemes, correlation structure, blocking and parametrization for the Gibbs sampler. *Journal of the Royal Statistical Society, Series B*, **59**, 291-317.

Roberts, G. O. and Smith, A. F. M. (1994) Simple conditions for the convergence of the Gibbs sampler and Metropolis-Hastings algorithms. *Stochastic Processes and their Applications*, **49**, 207-16.

Roberts, G. O. and Tweedie, R. L. (1994) Geometric convergence and central limit theorems, for multidimensional Hastings and Metropolis algorithms. Technical Report 94-9, University of Cambridge.

Rosenkranz, S. (1992) The Bayes factor for model evaluation in a hierarchical Poisson model for area counts, unpublished Ph. D. thesis, Biostatistics Department, University of Washington, USA.

Rosenthal, J. S. (1993) Rates of convergence for data augmentation on finite sample spaces. *Annals of Applied Probability*, **3**, 819-39.

Ross, S. (1996) *Stochastic Processes*, 2nd edn, Wiley, New York.

Rubin, D. R. (1987) A noniterative sampling/importance resampling alternative to the data augmentation algorithm for creating a few imputations when fractions of missing information are modest: the SIR algorithm, comment to a paper by Tanner and Wong. *Journal of the American Statistical Association*, **82**, 543-6.

Rubin, D. B. (1988) Using the SIR algorithm to simulate posterior distributions (with discussion), in *Bayesian Statistics 3* (eds J. M. Bernardo et al.), Oxford University Press, Oxford, pp. 395-402.

Rubinstein, R. Y. (1981) *Simulation and the Monte Carlo Method*, Wiley, New York.

Schmeiser, B. (1982) Batch size effects in the analysis of simulation output. Operation Research, **30**, 556-568.

Schmeiser, B. and Chen, M. H. (1991) General hit-and-run Monte Carlo sampling for evaluating multidimensional integrals. Technical Report, School of Industrial Engineering, Purdue University, USA.

Schruben, L. W., Singh, H. and Tierney, L. (1983) Optimal tests for initialization bias in simulation output. *Operations Research*, **31**, 1167-78.

Schwarz, G. (1978) Estimating the dimension of a model. *Annals of Statistics*, **6**, 461-4.

Shaw, J. E. H. (1988) Aspects of numerical integration and summarisation (with discussion), in *Bayesian Statistics 3* (eds J. M. Bernardo et al.), Oxford University Press, Oxford, pp. 411-28.

Shephard, N. (1994) Partial non-Gaussian state space. *Biometrika*, **81**, 115-31.

Shephard, N. and Pitt, M. K. (1997) Likelihood analysis of non-Gaussian measurement time series. To appear in *Biometrika*.

Silverman, B. W. (1986) *Density Estimation for Statistics and Data Analysis*, Chapman & Hall, London.

Singh, A. C. and Roberts, G. R. (1992) State space modelling of cross-classified time series of counts. *International Statistical Review*, **60**, 321-36.

Smith, A. F. M. (1984) Bayesian statistics. Present position and potential developments: some personal views (with discussion). *Journal of the Royal Statistical Society, Series A*, **147**, 245-59.

Smith, A. F. M. and Bernardo, J. M. (1994) *Bayesian Theory*, Wiley, New York.

Smith, A. F. M. and Gelfand, A. E. (1992) Bayesian statistics without tears: a sampling-resampling perspective. *American Statistician*, **46**, 84-8.

Smith, A. F. M., Skene, A. M., Shaw, J. E. H. and Naylor, J. C. (1987) Progress with numerical and graphical methods for practical Bayesian statistics. *The Statistician*, **36**, 75-82.

Souza, A. D. P. (1997) Approximate methods in dynamic Bayesian hierarchical models, unpublished Ph.D. Qualifying Exam, COPPE-UFRJ (in Portuguese).

Spiegelhalter, D. J., Thomas, A., Best, N. G. and Gilks, W. R. (1995a)

BUGS: Bayesian Inference Using Gibbs Sampling: Version 0.5. Technical report, Biostatistics Unit-MRC, Cambridge, UK.

Spiegelhalter, D. J., Thomas, A., Best, N. G. and Gilks, W. R. (1995b) *BUGS Examples: Version 0.5.* Technical report, Biostatistics Unit-MRC, Cambridge, UK.

Stephens, D. A. and Smith, A. F. M. (1992) Sampling-resampling techniques for the computation of posterior densities in normal mean problems. *Test,* **1**, 1-18.

Stone, M. (1974) Cross-validatory choice and assessment of statistical predictions (with discussion). *Journal of the Royal Statistical Society, Series B,* **36**, 111-47.

Swendsen, R. H. and Wang, J. S. (1987) Nonuniversal critical dynamics in Monte Carlo simulations. *Physics Review Letters,* **58**, 86-8.

Tachibana, V. (1995) Approximate methods in Bayesian models of randomized response and logistic regression, unpublished Ph.D. thesis, COPPE-UFRJ (in Portuguese).

Tanner, M. A. (1993) *Tools for Statistical Inference: Methods for the Exploration of Posterior Distributions and Likelihood Functions,* 2nd edn, Springer Verlag, New York.

Tanner, M. A. and Wong, W. (1987) The calculation of posterior distributions by data augmentation (with discussion). *Journal of the American Statistical Association,* **82**, 528-50.

Thisted, R. A. (1988) *Elements of Statistical Computing,* Chapman and Hall, New York.

Tierney, L. (1994) Markov chains for exploring posterior distributions (with discussion). *Annals of Statistics,* **22**, 1701-62.

Tierney, L. (1996) Introduction to general state-space Markov chain theory, in *Markov Chain Monte Carlo in Practice* (eds W. R. Gilks, S. Richardson and D. J. Spiegelhalter), Chapman & Hall, London, pp. 59-74.

Tierney, L. and Kadane, J.B. (1986) Accurate approximations for posterior moments and marginal densities. *Journal of the American Statistical Association,* **81**, 82-6.

Tierney, L., Kass, R. E. and Kadane, J.B. (1989) Fully exponential Laplace approximations for expectations and variances of nonpositive functions. *Journal of the American Statistical Association,* **84**, 710-6.

Vines, S. K., Gilks, W. R. and Wild, P. (1994) Fitting multiple random effects models. Technical report, Biostatistics Unit-MRC, Cambridge, UK.

Wakefield, J. C., Gelfand, A. E. and Smith, A. F. M. (1991) Efficient computation of random variates via the ratio-of-uniforms method. *Statistics and Computing,* **1**, 129-34.

West, M. (1985) Generalized linear models: outlier accommodation, scale parameters and prior distributions (with discussion), in *Bayesian Statistics 2* (eds. J. M. Bernardo et al.), North Holland, Amsterdam, 531-58.

West, M. (1992) Modelling with mixtures (with discussion), in *Bayesian*

Statistics 4 (eds J. M. Bernardo et al.), Oxford University Press, Oxford, pp. 503-24.

West, M. (1996) Some statistical issues in paleoclimatology (with discussion), in *Bayesian Statistics 5* (eds J. M. Bernardo et al.), Oxford University Press, Oxford, pp. 461-84.

West, M. and Harrison, P. J. (1997) *Bayesian Forecasting and Dynamic Models*, 2nd edn, Springer Verlag, New York.

West, M., Harrison, P. J. and Migon, H. S. (1985) Dynamic generalized linear models and Bayesian forecasting (with discussion). *Journal of the American Statistical Association*, **80**, 73-83.

Wild, P. and Gilks, W. R. (1993) AS 287. Adaptive rejection sampling from log-concave density functions. *Applied Statistics*, **42**, 701-9.

Zeger, S. L. and Karim, M. R. (1991) Generalized linear models with random effects: a Gibbs sampling approach. *Journal of the American Statistical Association*, **86**, 79-86.

Zellner, A. and Min, C. K. (1995) Gibbs sampler convergence criteria. *Journal of the American Statistical Association*, **90**, 921-7.

Author index

Subject index